T0136497

Life in 2030

The Sustainability and the Environment series provides a comprehensive, independent, and critical evaluation of environmental and sustainability issues affecting Canada and the world today.

**SUSTAINABILITY
AND THE
ENVIRONMENT**

John B. Robinson, David Biggs,
George Francis, Russel Legge, Sally Lerner,
D. Scott Slocombe, and Caroline Van Bers

Life in 2030
Exploring a Sustainable Future
for Canada

With an Introduction by Jon Tinker

A Project of the Sustainable Development Research Institute

UBCPress / Vancouver

Printed in Canada on acid-free paper ∞

ISBN 0-7748-0562-5 (hardcover)
ISBN 0-7748-0569-2 (paperback)
ISSN 1196-8575 (Sustainability and the Environment)

Canadian Cataloguing in Publication Data

Main entry under title:

Life in 2030

    (Sustainability and the Environment, ISSN 1196-8575)
    Includes bibliographical references and index.
    ISBN 0-7748-0562-5 (bound) – ISBN 0-7748-0569-2 (pbk.)

    1. Sustainable development – Canada. 2. Natural resources – Government policy – Canada. 3. Natural resources – Canada – Management. 4. Environmental policy – Canada. I. Robinson, John Bridger, 1953- II. Sustainable Development Research Institute. III. Series.

HC120.E5L53 1996   333.70971   C96-910022-1

UBC Press gratefully acknowledges the ongoing support to its publishing program from the Canada Council, the Province of British Columbia Cultural Services Branch, and the Department of Communications of the Government of Canada.

UBC Press
University of British Columbia
6344 Memorial Road
Vancouver, BC V6T 1Z2
(604) 822-3259
Fax: 1-800-668-0821
E-mail: orders@ubcpress.ubc.ca
http://www.ubcpress.ubc.ca

Contents

Figures

Acknowledgments

The Sustainable Society Project was a highly rewarding experience that will be fondly remembered by all project members. The authors would like to thank the Social Sciences and Humanities Research Council (SSHRCC) for the financial support that made this project possible; two grants were awarded in 1987 and 1991. Environment Canada also provided funding to continue research after our initial SSHRCC grant.

Thanks are also due to several people who provided valued input and support throughout the project. University of Waterloo graduate students Jeff Murdoch, Debbie MacFarlane, and Laura Kalbfleish joined the authors in the creation and exploration of the future scenario at the heart of this book. All members of our Advisory Committee (listed in Appendix A) provided support and encouragement throughout the project. For management of the project, thank you to Caroline Van Bers, project manager at the University of Waterloo, who kept us organized and honest. Finally, thanks to Deanna McLeod and Christine Massey, who coordinated the production of this manuscript after the project moved to the Sustainable Development Research Institute at the University of British Columbia.

Introduction

Jon Tinker

Until the late 1980s, environmentalists did not much like the future – or the present, for that matter. Ours was a backward-looking philosophy, a series of rearguard actions to protect a little wilderness from the chainsaw and the bulldozer, to set aside a few of our finest wetlands and mountains in national parks, to safeguard our seas from sewage and our air from automobile emissions, to conserve our fish stocks and our topsoil. We wanted to save what we could of our wild planetary heritage (and heritage *is* the past) from the unpredictable assaults of a frenzied, consumer-led technology.

Like Gandalf in Tolkien's *Lord of the Rings*, we felt that it was 'not our part to master all the tides of the world, but to do what is in us for the succour of those years wherein we are set, uprooting the evil in the fields that we know, so that those who live after may have clean earth to till. What weather they shall have is not ours to rule.'

This volume, the culmination of a five-year study led by John B. Robinson of the University of British Columbia, is more ambitious. It does try to rule tomorrow's weather. In a clear break from the environmentalism of Canute (hold back the tide) and Cassandra (it will end in tears), it spells out a coherent, practical, exciting, and above all optimistic vision of what twenty-first-century Canada could be like – if we decide to go there.

'I can call spirits from the vasty deep,' Shakespeare has Owen Glendower boast. 'Why, so can I, or so can any man,' replies Hotspur. 'But will they come when you call for them?' The spirits summoned by Robinson and his coworkers do materialize, helped by a rigorous analysis that disaggregates society into manageable components without losing sight of the whole, by a shrewd sense of realpolitik, and by the SERF computer-modelling system of Statistics Canada.

Life in 2030, as its subtitle suggests, is about sustainability. Its authors start from the rather dry definition of sustainable development adopted by the Brundtland Commission: to meet 'the needs of the present without compromising the ability of future generations to meet their own needs.'[1]

Robinson and his colleagues prefer to talk of working toward a 'sustainable society' rather than of sustainable development, and they approvingly quote Herman Daly to distinguish between 'growth' as a quantitative increase in the economy and 'development' as a qualitative improvement. They rightly insist that it would be hubris to claim we shall ever reach a sustainable society, though we can certainly aim for one that is more sustainable than we have at present.

Their frank admission that sustainability is not a wholly objective concept, and contains *desired* as well as *necessary* components, is reflected in the definition they finally adopt: 'the persistence over an apparently indefinite future of certain necessary and desired characteristics of the sociopolitical system and its natural environment.' The order of the last two terms is significant: Robinson and his colleagues see sustainability as deriving from ethical and political principles first and ecological ones second.

Politicians as well as environmentalists will learn from their discussion. The goal of sustainable societies, they argue, should not be to increase their *reliability* (resistance to breakdown) but to increase their *resilience* (capacity to recover from disturbance). 'The metaphor is that of safe-fail systems (which can fail "gracefully" without catastrophic repercussions), rather than fail-safe systems (which are less liable to break down initially but collapse entirely when breakdowns occur).' The distinction between reliability and resilience is drawn from ecology, which also suggests (worryingly, for a humanist) that intermittent catastrophes are necessary for long-term ecosystem health (wildfires in forestry, for example).

The reliability/resilience contrast is an ancient insight: one of Aesop's fables, first written down in the fourth century BC but certainly far older, concerns the reed that bent with the wind and survived the storm and the oak that did not. In the political field, a more recent statement of the same principle came in a 1981 House of Commons report on Canada's relations with Latin America: 'We must distinguish quite clearly between two different but frequently confused terms – the status quo and stability. Whereas the former frequently entails a rigid and repressive defense of existing social structures, the latter rests on consensus and the opportunity for change ... Canada has a special obligation and opportunity to stand on the side of the future.'[2]

This book is not a forecast of what its authors believe is likely to happen but a backcast from what they would like to see happen. Forecasting takes the trends of yesterday and today and projects mechanistically forward as if humankind were not an intelligent species with the capacity for individual and societal choice. Backcasting sets itself against such predestination and insists on free will, dreaming what tomorrow might be and determining how to get there from today. Forecasting is driving down the freeway and, from one's speed and direction, working out where one will be by nightfall. Backcasting is deciding first where one wants to sleep that night and then planning a day's drive that will get one there.

The society that relies on forecasters worries that it does not like where the freeway is going, but it usually ignores all the exit signs because it can visualize little other than business-as-usual. The society that listens to backcasters has to decide first where it wants to go. It can dream dreams, and it can work to make its dreams come true. Neither society, of course, can guarantee that it will arrive where it expects, but the latter will travel more hopefully, at the price of having to think.

As this book describes, the process of defining a coherent and feasible 2030 endpoint, and then constructing a convincing scenario to get from now to then, was not a one-off activity. Both the endpoint and the scenario went through many iterations, until Chapter 5 reached its present form: a 'future history' of Canada over the next thirty-five years, written as if its authors were living in 2030.

So where do Robinson and his colleagues think Canada should be in a generation or so? Their vision is detailed and complex. To understand it, one must read this book more than once, but a quick picture may be valuable at the outset.

Canada's population stabilized at thirty million in the year 2000, and it has remained at that level. Divorce is down slightly, fertility has remained at the 1990 level of about 1.7, and there are more non-traditional families. There has been a slight increase in life span, and the over-sixty-five age group has doubled from 11 to 22 per cent.

There is full employment, and the average formal work week is down to 27.5 hours. But almost everyone does voluntary work of some sort, and there has been a boom in private bartering and work for payment-in-kind. Many people work past retirement age. Either parent of a child up to three years old is paid a child-care wage, and all forms of social security have been merged into a guaranteed annual income.

Cities have become denser, with similar average living space. Urban sprawl has been halted, with suburbs converted into new urban nodes that have largely abolished commuting. Residential landscaping relies on native vegetation, requiring virtually no watering or chemicals.

The consumer society has been largely overturned. Although most people still have personal video and audio equipment, most other domestic durables are rented or borrowed from Common Goods Stores. Used clothing and shoe repair stores are common, and 'most people prefer to create their own styles imaginatively and inexpensively': real grunge rather than designer grunge?

Health care employs twice as many people as in 1990; alternative medicine is far more common, and most people are treated at home. Canadians eat less meat, sugar, and fat and more grain, fruit, and fibre; tobacco has virtually disappeared, and alcohol use is declining.

Education has similarly expanded, employing nearly 50 per cent more people. Half of the nineteen to twenty-four age group is in some form of postsecondary education, and adult education is much more common. Schools teach involvement in environmental management and decision-making from an early age.

Virtually everyone has a 'superbox' at home: a combined fax, computer, TV, and videophone – which can send as well as receive. Many TV programs are community-produced videos, and commercial TV helps to promote the environmental ethic.

Better communications have reduced the need for business travel, and there has been a small shift from air to rail and bus. Public transit has improved, with fast new rail links in major corridors such as Windsor-Montreal. Most households have one small electric car, for local use, and rent a larger vehicle, fuelled by either hydrogen or plant alcohol, for intercity travel. Cities and towns have been redesigned to encourage cycling, walking, and public transit.

Because travel is not subsidized, recreation has become more local. Small inns, restaurants, and picnic sites have flourished, at the expense of luxury hotels, because tax exemption has been withdrawn for business travel and entertainment.

Canada is still a food exporter, especially of beef and grain, but farmland has been reduced by 15 per cent, and agriculture has shifted toward lower input organic farming. Fruit and vegetables are more seasonal, especially since California's Imperial Valley ran out of water. Fish stocks are still recovering from the decline of the Pacific salmon fishery in the late 1990s, but domestic demand can still be satisfied.

Forest industries now employ more people, due to an increase in timber processing within Canada, but both production of and demand for lumber and paper have fallen. Houses are smaller and more durable. Mining has incorporated the concept of cradle-to-grave responsibility, and environmental damage has been greatly reduced. Sand and gravel extraction is down, with less need for roads. Iron and aluminum production is down too, with the reduction in the number of cars and the introduction of ceramic engine blocks; uranium production has stopped because, after the phaseout of nuclear power, it had no major markets.

Total energy production has fallen dramatically: coal by half (to reduce global warming and pollution), and petroleum and electricity by nearly 40 per cent, with natural gas use remaining constant. No new hydro schemes are being started.

Some of the biggest changes have been institutional, with a major devolution of power. The Canadian Charter of Rights and Freedoms has been joined by an Environmental Bill of Rights, by which every Canadian is guaranteed a healthy and sustainable environment and society; the Supreme Court will rule unconstitutional any legislation that conflicts with it. Under this bill, national goals and standards are set, but both the federal government and the provinces subcontract implementation to regional and local units (which are based mainly on ecological entities), provided standards are maintained. Each community develops and monitors its own sustainability program, and volunteers play a major role in implementation.

It would be easy to dismiss this scenario – set out in detail in Chapter 5 – as utopian science fantasy. But Robinson and his colleagues have accepted two key disciplines that fiction writers rarely follow. First, the scenario is internally consistent: energy production matches energy consumption, for example. Second, the scenario explains how Canada has moved from real-time 1990 to endpoint 2030: the reader must judge if this future history is feasible and credible.

Few Canadians will agree with all aspects of this possible future Canada. But most, I suspect, will be in sympathy with most of it. But what about the rest of the world? The future Canada outlined here implicitly assumes that most other countries have also moved heavily toward greater sustainability. Unless resource depletion and environmental degradation in the Caribbean and Central America, for example, are addressed soon, the present waves of emigrants from this region to North America are bound to grow. The lack of global as well as Canadian economic data does not necessarily invalidate the scenario. Indeed, the high level of grain, meat, timber, and mineral exports that Canada maintains in 2030 would, if much of the rest of the world is still on a business-as-usual track, mean far greater export earnings than today, as resource depletion elsewhere raises world market prices. And tourists from countries where natural beauty is largely gone, coming to enjoy Canada's relatively pristine environment, would be another major source of foreign exchange.

The Sustainable Society Project (SSP) was well under way when NAFTA came into force. That treaty places severe limitations on Canada's ability, alone, to reduce the exploitation of its minerals, petroleum, forests, and water in the way that this scenario outlines. If Canada has succeeded in doing this by 2030, the implication is either a major change in relations with the United States or a shift toward sustainability south of the border at least as significant as Robinson and his colleagues suggest for Canada. The present semipermeable frontier prevents even tobacco duty from being substantially different in the two countries. This volume does not claim to answer all such questions. Its value is in provoking them.

I began this preface by contrasting the essentially protectionist and defensive environmentalism that was prominent until the end of the 1980s with the more creative and optimistic outlook that is beginning to develop as ecology links up with future studies. There is another reason why this tendency is likely to continue: global warming. Because of the inertia of the planetary climate-control system, humankind is now committed to a substantial change in climate over the next thirty or so years. Canada, for example, is likely to see a lowering of the Great Lakes by up to a metre, a rise in sea level of perhaps more than a metre, less rainfall and more droughts in the Prairies, and much warmer winters in the North (leading to widespread reductions in permafrost).

Perhaps the most fundamental change will be that the areas where most current Canadian timber crop species grow will shift several hundred kilometres, generally north and east. Natural regeneration will not be able to keep pace with the rate of climatic change, and as trees die from warmer and drier conditions in the south, we will have to plant new forests farther north.

But a forest ecosystem is not just trees: it is made up of thousands of shrubs, herbs, mammals, birds, reptiles, insects, soil protozoa, and other organisms. Some of these species will be able to adapt to climatic change, and others will move northward. But a significant minority will not thrive in the more northern conditions of altered length of day, quality of soil, and other factors. The forest ecosystems that will evolve will not be the same ones shifted north: they will be new ecosystems, made up of a different mix of species. As for forest biomes, so too for wetlands, grasslands, rivers, seashores, and coastal waters.

Present-day wildlife conservation is based on protection and preservation of as many existing ecosystems as possible. Under global warming, many (maybe most) Canadian ecosystems will disappear in their present form, to be replaced by new systems made up of different species with different relations with one another.

For better or worse, ours is the last generation whose environmental and natural heritage philosophy can be based realistically on protecting some of the landscape that our parents knew. We will be forced to participate in designing our future, rather than struggling to preserve the best of our past.

In this volume, Robinson and his colleagues make an important contribution to enabling us to do this. Canada's future and the world's is not fixed by some consumerist, free market determinism; it does not have to be a continuation of every trend that holds today. We can plot our own destination and our own route, and we are not obliged to stay on the business-as-usual freeway.

Notes
1 World Commission on Environment and Development, *Our Common Future* (Oxford: Oxford University Press 1987).
2 Cited in J. Tinker, *On the Side of the Future: Development, Environment – and People* (London: Earthscan 1984).

Life in 2030

1

Exploring a Sustainable Future for Canada

John B. Robinson and D. Scott Slocombe

Background to the Study

As we move into the twenty-first century, an essential priority for Canadians, as for citizens of other countries, is the development of sustainable and environmentally benign patterns of resource utilization and socioeconomic development. The purpose of this book is to describe one attempt to investigate the feasibility and impacts of a future for Canada that is based on principles of environmental and sociopolitical sustainability.

From September 1988 through 1994, the Sustainable Society Project (SSP) explored the prospects for the development of a future Canadian society that is sustainable in environmental, economic, and social terms. Through scenario analysis, the project traced the path of Canada from the base year of 1990 forty years into the future, to approximately 2030. The project addressed sustainable futures at a national level. Sectorally, the scenario analysis focused on primary-resource sectors, the manufacturing sector, and the various consumption sectors. Chapter 2 discusses SSP objectives and methods in detail.

The Sustainable Society Project emerged out of the Canadian context of sustainable development initiatives and research, while also reflecting a more general interest in normative futures analysis intended to explore desirable rather than probable futures. The following discussion illustrates the SSP's methodological roots within futures analysis and modelling efforts.

Historical Context

The Sustainable Society Project has roots in efforts to anticipate or predict the environment and the future of human societies and economies. Ultimately, futures studies derive from the centuries-old utopian literature, but modern futures studies really began in the post-Second World War period with research related to national security and economic changes by consulting think-tanks such as RAND (Dickson 1972). Many consider the application of models to the future of environments and resources to have begun with the World Dynamics models of Jay Forrester et al. (Meadows et al. 1972). These models were the first widely known ones to look, in an admittedly highly simplified way, at the interaction of population, natural resources, and environmental quality. They tended to suggest major crises in terms of resource supply and human population.

While the 'Limits to Growth' and many other such models were, and still are, often controversial, they suggest the potential heuristic value of computers for simulating and predicting the future. Computer models have been widely developed since the late 1960s. Because computer predictions are often touted as more accurate than they really are, and because their predictions almost never come true, there have been backlashes and scepticism about them (Lee 1973; Skaburskis 1995). Others have sought to encourage the wiser use of computer models. This has included using them to explore alternatives and the potential impacts of different policies, in a qualitative way, rather than seeking or expecting precise, accurate predictions. Such models are often developed and used in participatory workshops where developing the multidisciplinary understanding to build them is as much the goal as the actual computer model. Several Canadians have had particularly strong roles here (Clark and Munn 1986; Holling 1978; Walters 1986). Computer models can also be integrated with non-computer-based futures methods.

Futures studies have involved a wide range of methods, including scenario development (Schwartz 1991; Wack 1985a,b), brainstorming and delphi/expert group approaches, cross-impact analyses, trend analyses, and others (Rotstein 1981). Since then, many more or less computer-model-based looks at the future of environments and economies have been undertaken (Hughes 1985). Methodologically and conceptually, the SSP is a part of these newer efforts to use models and other futures methods in an exploratory, heuristic, problem-solving way.

The conceptual origins of the Sustainable Society Project lie in the Conserver Society concept and research tradition in Canada. This was the first major project to combine sustainable development ideas and a futures orientation. The term 'Conserver Society' was coined in a 1973 report on resource policy by the Science Council of Canada, which urged that 'Canadians as individuals, and their governments, institutions and industries begin the transition from a consumer society preoccupied with resource exploitation to a conserver society engaged in more constructive endeavours' (Science Council of Canada 1973:39).

The Conserver Society concept was first examined in detail by GAMMA, a research group based in Montreal. In several reports published between 1975 and 1979 (Valaskakis et al. 1975, 1976, 1979), the GAMMA group identified and explored three successively more radical Conserver Society scenarios: expansion with efficiency ('doing more with less'), a stable industrial state ('doing the same with less'), and a 'Buddhist' scenario ('doing less with less').

The Conserver Society concept was also followed up by the Science Council itself, which published its major report on this topic in 1977, entitled 'Canada as a Conserver Society: Resource Uncertainties and the Need for New Technologies.' This report outlined five general policy thrusts: concern for the future; economy of design; diversity, flexibility, and responsibility; recognition of total costs; and respect for the regenerative capacity of the biosphere. It also contained a detailed discussion of the specific techniques and technologies that might be adopted in the transition to a Conserver Society. Generally, these differed little from the changes described in GAMMA's expansion-with-efficiency scenario. However, the Science Council chose not to discuss directly the sociopolitical dimensions of the Conserver Society argument: 'We have tried to stick to practical matters, with an incremental approach, to identify some of the technological paths that lead in the right direction, toward sustained relationships with material resources and the biosphere. Whether those paths, about which in our view we do not have much choice, imply other changes can only be decided by Canadians through democratic discussion' (Science Council of Canada 1977:14).

The importance of this emphasis on technological matters was that it allowed the Science Council to argue that significant improvements in emissions reduction, land-use and resource-development practices, environmental protection, and the efficiency of resource and materials use were all possible through improved technological development without significant reductions in material standards of living. This argument, which later became typical of a whole school of analysis in the energy and environmental fields, counterbalanced prevalent views that environmentalist approaches necessarily implied reduced living standards, and that such approaches amounted to resisting, rather than embracing, technological progress. On the other hand, such an approach precluded addressing other versions of the Conserver Society argument, such as GAMMA's second and third scenarios, which represented more significant changes in behaviour and development patterns.

Although the Science Council's Conserver Society report and related activities generated considerable interest both in Canada and abroad, this interest did not translate into any official political or policy response. Instead, the concept itself became part of the ongoing political debate about environmental issues and futures that occurred during the 1970s. Perhaps for this reason, the Science Council downgraded its efforts in this area. While it published a 1983 discussion paper on the Conserver Society (Schrecker 1983), no further research activity was undertaken, and the concept was allowed to die a natural death.

Despite this lack of official interest or follow-up, the Conserver Society concept had a significant influence on the development of environmentally based arguments in a number of areas, for example health (Hancock 1980) and agriculture (Hill 1985), as well as in providing a positive context within which to cast environmental arguments in general. Perhaps the most important area of application was in the energy field, which, since 1973, had been the subject of intense political debate and interest. In particular, the emergence of the 'soft energy path' concept, first articulated by Amory Lovins (1976a, 1977), became the vehicle for the most significant application of Conserver Society principles in Canada.[1]

Briefly put, the soft energy path argument suggests that demands for the services that energy provides (illumination, warmth, cooling, motive drive, mobility, etc.) can best be met through a combination of increased energy efficiency and renewable sources of energy that are diverse, flexible, and matched in geographic distribution and thermodynamic quality to end-use needs. This approach implies a massive reorientation of existing energy systems (and thinking) away from primary reliance on large-scale, centralized, and non-renewable energy sources toward systems characterized by high levels of energy efficiency and the use of small-scale renewables.

The soft energy path concept spawned a series of energy analyses, in various countries, intended to explore the technical and economic feasibility of soft energy futures. In Canada, these culminated in a province-by-province soft energy path study funded by the federal energy and environment departments and undertaken by Friends of the Earth Canada (1983/4; see also Robinson 1984). Using detailed scenario analysis, this study argued that under conditions of continued population and economic growth, it would be feasible and cost effective to operate the Canadian economy by 2025 using significantly less energy than in the base year of 1978 and supplying that energy largely through renewable sources of energy. However, by the time that the Friends of the Earth soft-path study was released in 1983 and 1984, the world oil glut and associated price decreases had caused the focus of federal energy policies in Canada to shift back to a traditional concern with oil and gas extraction and revenue issues. Although an updated version of the soft-path study was prepared and presented to the government (Torrie and Brooks 1988), the soft energy path concept in Canada, and elsewhere, appears to have suffered much the same fate in official circles as the Conserver Society concept.

With the reemergence of environmental issues in the mid-1980s and early 1990s, and the international activity and interest generated by the World Conservation Strategy, the Brundtland report, and the UN Conference on Environment and Development in Rio de Janeiro, many Conserver Society ideas resurfaced under the aegis of the sustainable development concept. In particular, concern over issues such as ozone depletion, global warming, and desertification are stimulating renewed interest in exploring environmentally benign patterns of resource use and economic activity. Most recently, the Second Assessment Report of the Intergovernmental Panel on Climate Change focused in part on the question of 'sustainable development' futures.

The Sustainable Society Project was an early part of this move to address the fundamental changes needed to achieve sustainability, based on an examination of the interconnections between resource use, economic activity, and social institutions and attitudes. The SSP also incorporated, and sought to demonstrate, methodological and theoretical lessons learned from the Conserver Society and other large-scale futures projects.

Methodological Background
From the 1970s experience with large, predictive models, it became increasingly apparent that traditional approaches to futures analysis and forecasting had a strong status quo bias and were essentially incapable of revealing or analyzing futures that represented significant departures from business-as-usual trends (Ascher 1978). Again stimulated by the early work of Lovins (1976a,b), energy analysts began undertaking 'backcasting' analyses, which postulated a different and apparently

more desirable future than that revealed by conventional forecasts, and assessing the feasibility and implications of such a policy path (Robinson 1982a). Others began to develop and use non-predictive methods of scenario analysis intended to alter the manner in which energy industry executives viewed the future (Wack 1985a,b). In a parallel development, researchers began to build energy 'end-use' models based on the depiction of physical processes rather than economic relationships. These models had the dual advantage of allowing non-predictive futures analysis of the backcasting or scenario-analysis type and permitting a much more detailed look at the potential for increased energy efficiency (Robinson 1982b). More recently, the backcasting and end-use modelling approaches have been generalized beyond energy issues to encompass questions related to global change and sustainable development (Robinson 1988, 1990, 1991).

Backcasting is a method of analyzing alternative futures. Its major distinguishing characteristic is a concern with how desirable futures can be attained. It is thus explicitly normative, involving working backward from a desired future endpoint or set of goals to the present, in order to determine the physical feasibility of that future and the policy measures that would be required to reach that point. In order to permit time for futures significantly different than the present to come about, endpoints are usually chosen for a time twenty-five to fifty years in the future.

Unlike predictive forecasts, backcasts are intended not to reveal what the future will likely be but to indicate the relative feasibility and implications of different policy goals. While the value and quality of a predictive forecast depend on the degree to which it accurately suggests what is likely to happen under specified conditions, backcasting is intended to suggest the implications of different futures, chosen not on the basis of their likelihood but on the basis of criteria external to the analysis (e.g., social or environmental desirability). No estimate of likelihood is possible because such likelihood would depend on whether the policy proposals resulting from the backcast were implemented. Thus, while the emphasis in forecasts is on discovering the underlying structural features of the world that would cause the future to come about, the emphasis in backcasts is on determining the freedom of action, in a policy sense, with respect to possible futures.

In order to undertake a backcasting analysis, future objectives are defined and then used to develop a future scenario. The scenario is specified by analyzing the technological and physical characteristics of a path that would lead toward the specified goals. The scenario is then evaluated in terms of its physical, technological, and socioeconomic feasibility and policy implications. Iteration of the scenario is usually required in order to resolve physical inconsistencies and to mitigate adverse economic, social, and environmental impacts that are revealed in the course of the analysis.

Because backcasting approaches explicitly introduce the question of policy choice, they serve to refocus the use of analysis away from responding to inevitable futures and toward exploring the nature and feasibility of alternative directions of policy. This helps to put the onus for choosing back where it belongs: in the policy arena.

The emergence of backcasting and end-use modelling approaches occurred within the context of a growing distrust of business-as-usual futures and a corresponding emphasis on openly normative, preferred futures. This is part of a more general critique of traditional, positive social science, which argues for more 'constructivist' approaches to the practice and evaluation of science and its use in the policy-making process (Robinson 1988, 1990, 1992). Recognizing the inescapably value-laden nature of all scientific analysis, such approaches try to make those values explicit and provide a means for their testing and evaluation, and for the identification of policy choices. This is particularly important with respect to what Salter (1988) has called 'mandated science,' that is, science undertaken in direct support of policy-making.

This approach to 'science for policy' suggests that the problem lies not so much with individual cases of bad analysis, or with improved methods of communicating science to policymakers or the public, as with a misconceived notion of what scientific analysis is and can provide to its users. This approach suggests that all mandated science is influenced strongly by embedded methodological and theoretical values. From this perspective, the goal of mandated science is not so much to find out true things as to make clear the points of difference, including value differences, between competing views of what the problem is. Put another way, the goal is to separate what is uncontroversial from what is not, and to make clear the underlying causes of the differences that exist. This raising of issues and clarification of assumptions and implications is a key goal of the SSP.

Moving Forward

The Sustainable Society Project took place against the backdrop of both a resurgence of interest in environmental issues and the emergence of new approaches to the study of desirable futures. It evolved from critiques of conventional wisdom and past efforts to move beyond them. The SSP focused on exploring the feasibility and impacts of explicitly normative, desired futures, within the framework of the perceived undesirability or riskiness of conventional, business-as-usual futures. The SSP's goals were to move beyond the energy focus of earlier Canadian scenarios and models of sustainability; to get past the consideration of technological fixes to examine lifestyle changes; and to go beyond scenarios of disaster to present images of a possible, positive future.

Discussions of sustainability are ripe for new approaches and ideas. The SSP is one new approach to identifying what sustainability might mean for Canada and some possible routes to getting there. There are, of course, many possible ways to move societies toward sustainability, and a key part of the project was the participants' learning about the process. Chapter 7 provides a retrospective on the project and some of the philosophical, methodological, and practical lessons learned from it.

Note
1 In fact, before publishing his seminal soft-path publications, Lovins published a paper in *Conserver Society Notes*, based on work undertaken for the Science Council, in which he sketched the outlines of a soft energy path future for Canada. Lovins called his two projections the Super Technical Fix model and the Conserver Society model (1976b).

References
Ascher, W. 1978. *Forecasting: An Appraisal for Policy-Makers and Planners*. Baltimore: Johns Hopkins University Press

Clark, W., and R.E. Munn, eds. 1986. *Ecologically Sustainable Development of the Biosphere*. New York: Cambridge University Press

Dickson, P. 1972. *Think Tanks*. New York: Ballantine Books

Friends of the Earth Canada. 1983/4. *2025: Soft Energy Futures for Canada*. 12 vols. Ottawa: Department of Energy, Mines and Resources

Hancock, T. 1980. 'The Conserver Society: Prospects for a Healthier Future.' *Canadian Family Physician* 26:320-1

Hill, S. 1985. 'Redesigning the Food System for Sustainability.' *Alternatives* 12 (3-4):32-6

Holling, C.S., ed. 1978. *Adaptive Environmental Assessment and Management*. Chichester: John Wiley & Sons

Hughes, B.B. 1985. *World Futures: A Critical Analysis of Alternatives*. Baltimore: Johns Hopkins University Press

Lee, D.B., Jr. 1973. 'Requiem for Large-Scale Models.' *Journal of American Institutional Planners* 34:163-78

Lovins, A. 1976a. 'Energy Strategy: The Road Not Taken?' *Foreign Affairs* (October):55-96

—. 1976b. 'Exploring Energy-Efficiency Futures for Canada.' *Conserver Society Notes* 1(4):3-16

—. 1977. *Soft Energy Paths.* New York: FOE/Ballinger

Meadows, D.H., D.L. Meadows, J. Randers, and W.W. Behrens. 1972. *The Limits to Growth.* Washington: Universe Books

Robinson, J.B. 1982a. 'Energy Backcasting: A Proposed Method of Policy Analysis.' *Energy Policy* 10:337-45

—. 1982b. 'Bottom-Up Methods and Low-Down Results: Changes in the Estimation of Future Energy Demands.' *Energy – The International Journal* 7:627-35

—, ed. 1984. 'A Soft Energy Path for Canada: Can It Be Made to Work?' Special Report on the Soft Energy Impacts and Implementation Workshop, 3-4 November 1983. *Alternatives* 12 (1):1-48

—. 1988. 'Unlearning and Backcasting: Rethinking Some of the Questions We Ask about the Future.' *Technological Forecasting and Social Change* 33:325-38

—. 1990. 'Futures under Glass: A Recipe for People Who Hate to Predict.' *Futures* 22:820-43

—. 1991. 'Modelling the Interactions between Human and Natural Systems.' *International Social Science Journal* 130:629-47

—. 1992. 'Of Maps and Territories: The Use and Abuse of Socio-Economic Modelling in Support of Decision-Making.' *Technological Forecasting and Social Change* 42 (4):147-64

Rotstein, E. 1981. *A Manual of Futures Forecasting Methods.* Toronto: Ontario Ministry of Transportation and Communications, Policy Planning and Research Division

Salter, L. 1988. *Mandated Science: Science and Scientists in the Making of Standards.* Boston: Kluwer Academic Publishers

Schrecker, T. 1983. 'The Conserver Society Revisited.' Ottawa: Science Council of Canada discussion paper

Schwartz, P. 1991. *The Art of the Long View: Planning for the Future in an Uncertain World.* New York: Doubleday

Science Council of Canada. 1973. *Natural Resource Policy in Canada.* Report No. 19. Ottawa: SCC

—. 1977. *Canada as a Conserver Society: Resource Uncertainties and the Need for New Technologies.* Report No. 27. Ottawa: SCC

Skaburskis, A. 1995. 'Resisting the Allure of Large Projection Models.' *Journal of Planning Education and Research* 14:191-202

Torrie, R., and D. Brooks. 1988. *2025: Soft Energy Futures for Canada – 1988 Update.* Report prepared for the Canadian Environmental Network Energy Caucus, for submission to the Energy Options Policy Review, Department of Energy, Mines and Resources Canada. Ottawa

Valaskakis, K., P.S. Sindell, and J.G. Smith. 1975. *Tentative Blueprints for a Conserver Society in Canada.* Montreal: GAMMA

—. 1976. *Conserver Society Project – Phase II Report.* 4 vols. Montreal: GAMMA

Valaskakis, K., P.S. Sindell, J.G. Smith, and I. Fitzpatrick-Martin. 1979. *The Conserver Society: A Workable Alternative for the Future.* New York: Harper and Row

Wack, P. 1985a. 'Scenarios: Uncharted Waters Ahead.' *Harvard Business Review* 5:72-89
—. 1985b. 'Scenarios: Shooting the Rapids.' *Harvard Business Review* 6:139-50
Walters, C. 1986. *Adaptive Management of Renewable Resources*. New York: Macmillan

2
The Sustainable Society Project
John B. Robinson, Caroline Van Bers, and David Biggs

We can approach the future in many ways. The conventional route is to project forward from present trends and make adjustments to avoid some of the uglier prospects. But such an approach does not allow for the need to make more significant changes. When basic threats to survival and well-being loom large, mere adjustments to the present path seem unlikely to be sufficient. Instead, we need to imagine more desirable futures and explore what they would be like and how they could be attained. This chapter describes how we undertook such an exploration in the Sustainable Society Project (SSP).

The Genesis of the Project

The Sustainable Society Project grew directly out of the substantive and methodological critiques of conventional wisdom about the future described in Chapter 1. The initial impetus for the project was the recognition that studies of alternative energy futures (soft energy paths) were inadequate because they did not address the non-energy parts of society. It was hard to make a 'soft energy path' work in a society that was otherwise a business-as-usual world. More importantly, presumably the same kinds of environmental and social reasons that suggested the desirability of alternative energy paths would also apply to other sectors of society. In other words, what was needed was a soft-path kind of analysis applied to the whole of Canadian society. To our knowledge, no one had ever attempted this kind of work.

Second, most of the studies that did attempt to project alternative, more environmentally benign futures limited themselves to 'technical fix' changes. That is, they assumed that lifestyles would not change significantly, that people would still live and consume much as they do now, and then simply studied the potential to produce and consume goods more efficiently. In the SSP, we were also interested in exploring what might be the effects of significant changes in lifestyle, in what was consumed and in how people lived.

A third major reason for undertaking the project was the feeling that too much of the environmental literature about the future was characterized by what might be called an apocalyptic vision, or a gloom-and-doom approach. Much of the literature projected futures that demonstrated various forms of environmentally related collapse, or at least offered fairly dire predictions of the consequences of continued business-as-usual approaches.[1] While we shared these concerns, we believed that, beyond a certain point, a continuous diet of such gloomy projections was likely to induce apathy, or worse. If people come to believe that the world is rapidly descending into environmental collapse, that we have already exceeded planetary carrying capacity, then many may believe that they should get what they can while it is still there to be got. Our view was that there was and is a need for more images of desirable futures, and that we therefore needed to try to articulate and test the feasibility of such positive images. Thus, the SSP was intended to extend past work beyond energy issues in order to provide a comprehensive look at an overall sustainable society and, in so doing, to explore and articulate the implications of a future chosen on the basis not of likelihood but of environmental and social desirability.

The origins of the Sustainable Society Project can be traced to a workshop held at the University of Waterloo in the fall of 1983 (Robinson 1984). The workshop was funded by the Social Sciences and Humanities Research Council of Canada (SSHRCC) under the Human Context of Science and Technology Program. Its purpose was to explore the questions about implementation, and social, economic, and environmental impacts, that had been raised by the recently completed Friends of the Earth soft energy path study (1983-4). That study had focused on the technical and economic potential for soft energy technologies, but had also pointed to the need for further work on implementation and impact.

One of the conclusions reached at the workshop was that there was a need to delineate more carefully the economic structure and resource requirements of the kind of society that would both support a soft energy path and be relatively environmentally benign. It was determined that a research project intended to outline the resource requirements and socioeconomic impacts of what later came to be called a 'sustainable society' scenario for Canada would represent the next logical step for research.

The idea for such a study was made possible by the development and availability of a set of long-term simulation models at the University of Waterloo. Originally developed at Statistics Canada, the Socio-Economic Resource Framework (SERF) – an earlier version of which had been used to develop economic scenarios for the Friends of the Earth soft energy path study mentioned above – was installed in the Faculty of Environmental Studies in the fall of 1984. As discussed in more detail below, SERF, which represents a globally unique integration of socio-economic data and modelling capability, was designed to permit long-term normative simulation analysis of the kind envisaged in the proposed study.

During 1985 and 1986, a project proposal was developed and refined and a research team assembled. The project was awarded funding from SSHRCC in the fall of 1987, and work began in the fall of 1988. Final model runs and analysis were completed in 1994.

The interdisciplinary nature of the project required a team approach. Assembled was a core project team of five faculty with expertise in modelling and futures studies, sustainable development, ecology, resource management, sociology, religion, and values. A number of graduate students were involved in the project, and three completed theses on specific sectors; the theses then became the basis for scenario inputs. The project also sought occasional contributions from other faculty with expertise in parks and forestry, water resources and fisheries, environmental politics, and northern development. An advisory committee was formed, consisting of twenty-six individuals who collectively represented much of the Canadian expertise on environmentally benign futures.

Purpose and Objectives

A sustainable society has sociopolitical as well as environmental and technological implications. The challenge is to analyze the implications of a sustainable society in an integrated way that accounts for both of these dimensions. Moreover, the human and technological dimensions of a socioeconomic system are dynamically interrelated; sociopolitical structures are reflected in technological and economic development, and vice versa.

The overall purposes of this project were to create a scenario, based on principles of sustainability, of the future development of Canadian society; to assess this scenario in terms of its feasibility, implications, and implementation requirements; and to contribute to the development of a network of groups and individuals interested in sustainable futures for Canada.

The Sustainable Society Project was intended to go beyond previous research in Canada in several interconnected ways. First, previous work tended to be distinguished by comprehensive but qualitative treatment (Conserver Society analysis) or by rigorous and quantitative scenario analysis of one policy area (soft energy path studies). The SSP provided rigorous scenario analysis of all sectors of the economy. Second, the project was not limited to technical fix solutions but included analysis of lifestyle and social change. Finally, the SSP contained a detailed analysis of the links between sustainability values and socioeconomic and technological change and development.

While the analysis performed in the project involved multiple iterations to produce a balanced scenario, the project was designed to test the feasibility and impacts of a single sustainable society scenario. Although analysis of several alternative versions of a sustainable society, and comparison to at least one business-as-usual future, would have been desirable, such an approach was precluded by time and resource constraints.

The general approach used to integrate the various dimensions of sustainability in the project was to define sustainability as a normative ethical principle having both sociopolitical and environmental/ecological dimensions, and then to develop scenario design criteria in each of these two areas that could be used to drive the scenario analysis. The intention was to iterate through as many attempts at developing such scenarios as necessary to produce a reasonably consistent picture that conformed to the initial design criteria, and then to assess the feasibility, implications, and implementation requirements of that scenario.

The specific objectives of the project were:

(1) to develop sociopolitical and environmental/ecological design criteria for the scenario analysis that are based on sustainability conceived as a normative ethical principle
(2) to create quantitative subscenarios of technological and economic development in SERF based on these design criteria and, through an

iterative process, to integrate them into one physically consistent scenario of a sustainable future

(3) to determine and then evaluate the general social, economic, environmental, and political implications of this final scenario, assess its overall feasibility, and analyze the sociopolitical implementation measures required for it to occur

(4) to contribute to the development of a network of interested groups and individuals in the field through involvement in the project

(5) to produce and disseminate a final report summarizing the results of the analysis.

Methods

Scope

The project addressed sustainable futures at a national level. The spatial scope of the project was therefore the whole of Canada. International trade was addressed explicitly through the use of trade and balance-of-trade calculations.

Temporally, the scenario analysis extended forty years into the future from the base year of 1990, to approximately 2030, allowing sufficient time for the complete turnover of capital stocks, the generation of new structural relationships in the economy, and the development of new institutional relationships in the political system. Comprehensive descriptions of the physical and technological state of the system are available for each year of the scenario evolution.

Sectorally, the scenario analysis focused on the primary-resource sectors (energy, forestry, mining, agriculture, fisheries, and water), the manufacturing sector, and the various consumption sectors (dwellings, consumer goods, health, education, transportation, retail trade, office buildings, etc.). Subscenarios of the sectoral subject areas were developed and then integrated into an aggregate scenario.

The impact and implementation analysis involved investigation of the consistency of the sustainable society scenario with project design criteria and assessment of the changing institutional relationships and political structures that might be expected to result from, and help to cause, the technological and economic developments described in the quantitative scenarios. This involved an assessment of the compatibility of these evolving sociopolitical forms with each other, with principles of environmental sustainability, and with value principles basic to Canadian society.

Theoretical Approach

Environmental problems pose important questions not only about scientific knowledge and technological change but also about social and political organization. As discussed above, this project was based on the view that science and technology are deeply value-laden. Therefore, it was necessary to examine closely the relationship between technology and social organization and between science and its use in decision-making; it was also necessary to develop technologies, and systems of expertise, that are overtly grounded in the values that technology and expertise are intended to serve (Winner 1986).

The result was an openly normative approach to the assessment of technology and social organization. In this project, that approach was represented by the linkage between technological development, social change, and sustainability values, and by the development of a scenario-analysis method that could be used to express and evaluate different values and views of preferred futures. On this view, our role as analysts was not to determine what is the right course of action but to propose both our preferred alternative to current policy (which is explicitly grounded in a set of values) and a means by which others could propose and assess their own views and preferences.

The values forming the normative basis for this project were a version of those associated with the terms 'sustainable society' and 'sustainable development' (Brown 1981; Clark and Munn 1986; World Commission on Environment and Development 1987). As discussed in Chapter 1, they imply a critical, or even radical, approach to technological and social development. In order to assess the implications of these values and to implement the normative approach to science, technology, and analysis described above, it was necessary to use an analytical approach that departs from the predictive and ostensibly value-free orientation typical of most futures studies (Robinson 1988). This project used the 'backcasting' method of scenario analysis discussed in Chapter 1. Backcasting involves defining goals, articulating them in terms of preferred future states of the system being analyzed, and then attempting to construct a path of technological and social development between the present and the desired endpoint (Robinson 1990). The goal is to assess the feasibility and impacts of normatively defined futures. In so doing, the analysis serves as a kind of consistency and feasibility check on the values in terms of which the normatively defined futures are constructed.

A unique characteristic of this project is its use of the SERF modelling system for scenario generation and analysis. SERF is an implementation of the design approach to socioeconomic modelling (Gault et al. 1987). Design approach models are intended to examine the physical feasibility of alternative policy goals over the long term (thirty-five to seventy years). They are thus not useful for predicting future events; they are intended, rather, to be used for backcasting analyses.

A Description of SERF
SERF is a modelling framework built at the Structural Analysis Division of Statistics Canada (Hoffman 1986; Hoffman and McInnis 1988). (See Appendix B for a more detailed description of the SERF system.) The models in SERF are not 'behavioural'; that is, they represent physical accounting frameworks that show the implications (i.e., the physical consistency) of assumptions about future behaviour chosen by the user. This means that SERF does not include prices, or other economic variables, inside the modelling system. SERF does not calculate optimal or equilibrium outcomes, but simply shows the physical consequences of making certain behavioural choices. This aspect makes it particularly suitable for exploring the physical feasibility, rather than the likelihood, of alternative scenarios.[2]

SERF is organized into four major components: demography, consumption, production, and resource extraction. The user supplies input scenarios in the form of assumed future values for the approximately 1,700 multidimensional SERF variables, and SERF combines these time series inputs into integrated scenarios and assesses the physical consistency, over time, of the resultant overall scenario of the evolution of Canadian society. SERF keeps track of the flows of energy, materials, goods, and labour throughout the simulation. The disaggregated and comprehensive nature of SERF allows the user to undertake detailed analysis of changing efficiencies, technological substitutions, labour productivity, and so on.

In SERF, the demography component (population, household formation, and labour force) begins the simulation process by calculating the future population, household, and labour force levels as determined by the input assumptions. These numbers are then used to drive a set of consumption calculators (housing, consumer goods, health care, education, transportation, offices, communications, and retail trade) to determine the amount of goods and services required. Each of these consumption calculators, of course, contains many additional input assumptions which determine the kind and quantity of goods and services required. The required goods are then 'produced' in the production component using an input/output table.

Finally, resource production calculations are performed. Inputs to the five natural resource production sectors (agriculture, energy, forestry, mining, and fisheries) are specified as planned capacity and operating characteristics for those sectors, rather than being determined directly by the level of demand for the products of those sectors. The level of production resulting from the input assumptions for each of these sectors is compared in SERF to the level of demand for those products in the rest of the economy. They may or may not match. For example, the amount of lumber produced in the SERF scenario is a function of planned capacity and operating characteristics of the forestry and lumber industries. Elsewhere in SERF, the demand for lumber from all other sectors in the economy is added up and then compared to this supply. Any mismatch is expressed as a 'tension,' which must be resolved by changing some of the inputs and re-running SERF in an iterative process.

While the existence of SERF allowed the Sustainable Society Project to undertake detailed quantitative scenario analysis of a kind not hitherto possible, SERF does not encompass all dimensions of Canadian society relevant to this project enquiry. In addition to the lack of economic analysis mentioned earlier, SERF does not describe the decision-making processes and changing institutional arrangements that are associated with the technological/physical scenarios that it generates. Therefore, as discussed below, analysis of sociopolitical transformation was developed outside SERF but connected to the scenarios generated in SERF.

Organization of the Project

In 1987, the Sustainable Society Project team was awarded $100,000 from the Social Sciences and Humanities Research Council of Canada (SSHRCC). In September 1990, the project received a $25,000 contract from Environment Canada to continue research; in March 1991, it was awarded $110,000 from SSHRCC for further work. The research team consisted of five analysts at the University of Waterloo and Wilfrid Laurier University, a graduate student project manager, and a number of graduate and undergraduate students undertaking theses within the project, as well as a larger group of project researchers, an advisory committee, and a network of groups and individuals with an interest in the concept.

The general approach of the project was to articulate sustainability values, derive scenario design criteria based on these values, develop a qualitative picture of Canada in 2030 consistent with these criteria, construct a quantitative scenario in the computer program intended to describe a path between 1981 and 2030 that takes us to that future, and analyze the implications, feasibility, and implementation requirements of that scenario.

The first step was to articulate and elaborate the values and principles underlying the sustainability scenario. Sustainability was defined as a normative ethical principle having both environmental/ecological and sociopolitical dimensions. On that basis, several basic value principles and a set of more specific ecological and social principles were articulated. These principles were used to generate the environmental/ecological and sociopolitical scenario design criteria, and the description of economic, legal, political, and individual decision-making in Canada in 2030. Detailed qualitative and quantitative descriptions for each sector of society over the period 1981 to 2030 were created, based on the principles and scenario design criteria. These descriptions were then used to develop inputs to the SERF modelling system.

This computer software was designed to allow the simulation of alternative futures by modelling physical aspects of the Canadian socioeconomic system. This computer analysis described energy, labour, and material flows in the four major sectors: demography, the various consumption sectors, the primary-resource sectors, and manufacturing. Based on sustainability principles, smaller, more specific scenarios of individual subject areas were developed, and these subscenario design criteria were then integrated into an overall scenario.

The computer model then added up and compared the activities and assumptions in all sectors, and identified discrepancies between supply and demand for goods, labour, services, and materials. Areas of social, economic, or physical unsustainability in the scenario were thus identified by the computer, and on the basis of these inconsistencies, the specific story-line assumptions that had been entered were reevaluated in order to mitigate problem areas. The subscenarios and the overall scenario were rerun as many times as required to resolve physical inconsistencies and to produce a balanced scenario consistent with the project's initial design criteria. The sociopolitical analysis then involved investigation of the changing institutional relationships and political structures that might be expected to accompany the technological and economic developments described in the computer-generated scenario.

After creation of the scenario, the next steps in the project involved considering the longer term policy options and changes in individual and organizational behaviour that would be associated with successful implementation of the project scenario. In May 1992, preliminary results were presented to the SSP Advisory Committee, which established four working groups to pursue issues related to implementation and outreach, covering policy, media/education, future history, and ethics and distribution. A more developed scenario was then presented to civil servants in Ottawa in October 1992, at a meeting hosted by the National Round Table on the Environment and the Economy. Subsequent work at UBC in 1993, 1994, and 1995 involved reiterating the scenario to ensure internal physical consistency and compatibility with sustainability values, undertaking more detailed analysis of the lifestyle changes associated with the energy analysis, and writing up the results.

Outreach and Products

The Sustainable Society Project contributed to the ongoing debate about environmentally benign futures in Canada in several ways. First, the project refined, integrated, and tested some of the concepts and proposals that have emerged out of the Conserver Society environmental and sustainability arguments in Canada. This work involved examination of a number of the specific proposals that have been developed in support of sustainable development approaches. For example, in our analysis, soft energy systems were combined with sustainable agriculture proposals, alternative health care approaches, and so on. The use of SERF and the associated external analysis allowed these proposals to be considered together in an integrated way and to be evaluated in terms of feasibility and implications. The result of the scenario analysis thus gives some indication of whether and how an overall sustainable society can be made to hang together.

Second, an important component of our goal to advance the debate in Canada consisted of fostering the formation and maintenance of a network of groups and individuals interested in environmentally benign futures for Canada. We did this by maintaining contact with groups and individuals who are interested in the project. This contact occurred through both the Advisory Committee and direct mail. In this way, we hoped that the project could act as a catalyst for new ideas and the further development of sustainability arguments.

Finally, the results of the project itself were intended to provide information that can be used as a lobbying tool in ongoing political debates about desirable futures for Canada. In this respect, the focus of the project on developing and testing positive images of future development was important. Environmental arguments are often perceived as being essentially negative, taking the form of critiques of existing or proposed developments. This project was intended to develop and present a positive alternative development path for Canada. To the extent that it succeeded, some of the debate can be shifted from the clamour of competing negative claims to a focus on the relative desirability of alternative images of the future.

In support of these various goals, a number of reports and outreach activities were undertaken. An SSP newsletter was created and distributed to several hundred people and groups over the course of the project, and a number of conference presentations were made and workshops held. Eight working papers were completed and distributed to anyone requesting them, and a number of published papers and student theses were produced. (See Appendix A for a list of project participants, workshops, and papers.)

Conclusions

The Sustainable Society Project was an ambitious attempt to tie together some of the diverse threads of the environmental arguments that have developed from the Conserver Society and sustainability tradition in Canada. In so doing, it represented an application of a set of methodological approaches and principles that have emerged out of the same tradition. It was, in fact, the perceived need to develop methods of analysis oriented toward the examination of unconventional futures that led environmentalists and soft energy analysts to develop backcasting approaches, and Statistics Canada researchers to develop SERF. The SSP combined these methodological and substantive approaches in an integrated analysis.

The analysis summarized in this book can provide only a preliminary assessment of the feasibility and impacts of a sustainable future for Canada. In so doing, however, it can contribute to the ongoing debate about such futures. Of course, no futures analysis of the type proposed here can be definitive or exhaustive. At best, it can reveal the rough outlines of a desirable future, and suggest that such a future seems feasible and therefore worth striving for.

This does not mean that the analysis and analysts have no role in the political debate about desirable futures. It means that futures analysis should not substitute for decision-making. The role is to explore various versions of possible futures, and to indicate their apparent feasibility and implications. The rest must be left, as it should be, to the political process.

Notes

1 This tradition has continued from the early days of *The Limits to Growth* (Meadows et al. 1972). For a more recent example, see Rees and Wackernagel (1994).
2 For our purposes, the exclusion of economic relationships from the model is desirable, in that it allows unconstrained analysis of behavioural and technological change. That is, because we wanted to explore patterns of technological change and behaviour that differ significantly from past experience, we did not want to 'hard-wire' behavioural relationships among, say, income and energy use into the model. This freedom, however, comes at the cost of being able to analyze the macroeconomic dimensions of our scenario.

References

Brown, L. 1981. *Building a Sustainable Society*. New York: Norton
Clark, W., and R.E. Munn, eds. 1986. *Ecologically Sustainable Development of the Biosphere*. New York: Cambridge University Press
Friends of the Earth Canada. 1983/4. *2025: Soft Energy Futures for Canada*. 12 vols. Ottawa: Department of Energy, Mines and Resources
Gault, F., K.E. Hamilton, R.B Hoffman, and B.C. McInnis. 1987. 'The Design Approach to Socio-Economic Forecasting.' *Futures* 19 (1):3-25
Hoffman, R. 1986. 'Overview of the Socio-Economic Resource Framework (SERF).' Working Paper 86-03-01, Structural Analysis Division, Statistics Canada, Ottawa
Hoffman, R., and B. McInnis. 1988. 'The Evolution of Socio-Economic Modelling in Canada.' *Technological Forecasting and Social Change* 33 (4):311-24
Meadows, D.H., D.L. Meadows, J. Randers, and W.W. Behrens. 1972. *The Limits to Growth*. Washington: Universe Books
Rees, W.E., and M. Wackernagel. 1994. 'Ecological Footprints and Appropriated Carrying Capacity: Measuring the Natural Capital Requirements of the Human Economy.' In *Investing in Natural Capital: The Ecological Economics Approach to Sustainability*. Ed. A.-M. Jansson, H. Hammer, C. Folke, and R. Costanza. Washington, DC: Island Press
Robinson, J.B., ed. 1984. 'A Soft Energy Path for Canada: Can It Be Made to Work?' Special Report on the Soft Energy Impacts and Implementation Workshop, 3-4 November 1983. *Alternatives* 12 (1):1-48
—. 1988. 'Unlearning and Backcasting: Rethinking Some of the Questions We Ask about the Future.' *Technological Forecasting and Social Change* 33:325-38

—. 1990. 'Futures under Glass: A Recipe for People Who Hate to Predict.' *Futures* 22:820-43

Winner, L. 1986. *The Whale and the Reactor: A Search for Limits in an Age of High Technology*. Chicago: University of Chicago Press

World Commission on Environment and Development. 1987. *Our Common Future*. Oxford: Oxford University Press

3
Defining a Sustainable Society
John B. Robinson, George Francis, Sally Lerner,
and Russel Legge

Introduction

This chapter provides a working definition of sustainability that can be used to describe a sustainable Canadian society. A set of general princi- ples of sustainability is articulated, and it forms the basis for the discus- sion in subsequent chapters. Chapter 4 describes the scenario design criteria used to translate the values and principles articulated in this chapter into inputs to the sustainability scenario. The results of the sce- nario analysis are set out in Chapter 5, and some of the policy and eco- nomic implications of the scenario are explored in Chapter 6.

The Concept of Sustainability

Over the past quarter century, there has been a growing recognition of the harmful effects and economic costs of continued environmental degradation.[1] Official recognition of these problems is reflected in the establishment of environmental agencies, ministries, and other organi- zations, the growth of environmental policies and regulations in both the public and private sectors, and the growing political profile of envi- ronmental issues.

Initially, concern with environmental issues focused mainly on short-term and local effects. More recent analysis has concluded that human activities may be causing significant effects on global biogeo- chemical and biogeophysical systems, with potentially large-scale and long-term impacts.[2] The best-known examples of these global effects are climatic warming due to greenhouse gases and ozone depletion due to the emission of CFCs (IPCC 1992). The extent of the potential im- pacts has led scientists to suggest that we are conducting what amounts to a global experiment on the biosphere, with potentially fatal conse- quences for humanity (Environment Canada 1989).

In order to address such problems, we need to consider the social causes and impacts of the human activities that underlie them. This consideration, in turn, raises some important questions about the fundamental character of our society. For example, it has been argued that environmentally destructive human activities are based on behaviour that is deeply rooted in modern industrial civilization's underlying assumptions and beliefs about nature and humanity's place in it (Naess 1973; Capra 1983; Berman 1984; Devall and Sessions 1985; Suzuki 1994; McLaughlin 1993). Others see the negative environmental consequences of such activities as connected to other negative social characteristics of modern society, such as poverty or social alienation, and argue that the exploitation of nature by human beings is closely connected to the exploitation of humans by other humans (Leiss 1974; Bookchin 1986). It is therefore necessary to consider the social attitudes and practices that give rise to environmental (and social) impacts.

From this perspective, the concept of sustainability goes beyond the continued existence of biophysical life-support systems. Because of these broader questions about the cultural roots of human activities and their social consequences, sustainability issues have sociopolitical dimensions as well. They have to do not simply with the alteration of certain social practices that give rise to unpleasant environmental consequences but also with the underlying causes of those practices, and with the linkage between sociopolitical and environmental states. In other words, a sustainable society must be sustainable in both environmental and sociopolitical terms.

This was also the conclusion reached by the World Commission on Environment and Development (1987:8), commonly known as the Brundtland Commission. According to the commission,

> Humanity has the ability to make development sustainable – to ensure that it meets the needs of the present without compromising the ability of future generations to meet their own needs. The concept of sustainable development does imply limits – not absolute limits but limitations imposed by the present state of technology and social organization on environmental resources and by the ability of the biosphere to absorb the effect of human activities.

While adopting this orientation of the Brundtland Commission, we recognize some of the concerns about the concept of sustainable development expressed from within the environmental movement:[3]

To some in the environmental movement, [the term 'sustainable development'] sounds positively sinister. Their fears are that the phrase will become a code for legitimizing all development so long as there is some token environmental scrutiny of it. But for all environmentalists, including myself, that approach is simply not good enough. It is a continuation of the route that brought us to our present difficulties. A far more pervasive approach to incorporating environmental values into routine economic decision-making is needed. (Holtz 1988)

The underlying concern expressed is the degree to which the sustainable development argument simply papers over serious differences between pro-growth and anti-growth arguments and deflects attention from the real sociopolitical and economic changes required.[4] We recognize that ecological and societal sustainability may well imply patterns of economic and social development that are quite different from those implied in traditional views of continued economic development. The challenge is to articulate and test the degree to which such changes are required if we are serious about the concept of sustainability.

Sustainability as an Ethical Principle

While the practical reasons for concerns about sustainability are compelling, they do not by themselves provide an adequate basis for developing design criteria for a sustainable society. Although the concept of sustainability has been discussed at length (Clark and Munn 1986; Barbier 1987; Brown et al. 1987; World Commission on Environment and Development 1987; United Nations Conference on Environment and Development 1992; World Resources Institute 1994), there has been relatively little discussion of the underlying ethical principles of the concept. We begin with two observations about the relationship between human beings and the natural world.[5]

First, the survival of human beings depends utterly on the environment for air, water, food, and material resources, and to serve as a sink for their wastes. Its continued existence is a prerequisite for the existence of human society. Second, humanity does not understand much about the complex sets of interactions and processes of the biophysical world around us. That is, a significant degree of ignorance about the external world is characteristic of the human condition. This ignorance suggests that a basic caution is in order when interfering with natural processes.

These two observations provide a practical, albeit anthropocentric, justification for concern about sustainability. However, we wish to combine this justification with an assertion of the ethical principle that the existence of the natural world is inherently good. That is, the natural world and its component life forms, including humanity, have value by and for themselves. (This belief, in turn, is based on the assumption that affirming the intrinsic value of the natural world and of humanity is not self-contradictory.) Such a principle provides an ethical basis for valuing sustainability and for affirming autonomy in nature.

This principle, which is present in some form in many cultures and religions, is a fundamental ethical principle that need not be justified in terms of any other value. For our purposes, however, it is based on the assumption that life evolves something like the way suggested by current evolutionary theory, which suggests that there is a process of selection through which the natural world determines what is sustained. This selection follows a dynamic pattern, which in itself is a continuing affirmation of our ethical principle. With respect to sustainability, this means that the environmental/ecological sphere must be able to regenerate itself through natural evolution. And if it does, our ethical principle suggests that it will thereby be able to contribute in its own creative ways to the enrichment of life as a whole. Thus, continuous regeneration of the natural world becomes a basic value.

The assertion that the continued existence of a world and the life on it is a good thing does not, however, tell us enough about what kind of things should be sustained and where, or about how we should behave. For example, it is not clear a priori that any component of the biosphere, such as a specific species, forest, or ecosystem (let alone any particular social practice, such as parliamentary democracy or market economics), should be valued in itself. However, it does imply that those characteristics of natural systems that are the result of biophysical evolutionary process are valued characteristics of the natural world. We then have to rely on ecological science in order to obtain the best current understanding of natural biophysical process. We will make use of this approach in our discussion of the principles of environmental/ecological sustainability below.

Considering the ethical underpinnings of humanity's treatment of the natural world is not enough, however. In keeping with the general conceptualization of sustainability introduced above, we also need to discuss the ethical dimensions of sustainability as they relate to sociopolitical issues.[6] To do so, we begin with another principle: nature's gift of self-consciousness is, as far as we know, the trait that distinguishes humans from other animal forms. It is the source of all concepts of the relationship between humanity and the natural world, of all technologies, and of moral responsibility. We combine that principle with the observation that the development of modern industrial society presupposes the existence of a type of science and technology that is historically unique, and uniquely capable of transforming the natural world and human capabilities. Thus, we need to be concerned with the moral dimensions of sustainability, in terms of the particular form that human self-consciousness takes, as expressed in sociopolitical organization and human behaviour.

Such self-consciousness is gained by the capacity of the mind to work symbolically, that is, to represent to itself alternative states of existence and to create different 'universes of meaning' (e.g., ideologies, religions, and worldviews). Sociopolitical systems are universes of meaning that structure the evolution of individual and collective practices, and the institutions that make such practices possible. A successful universe of meaning has *cultural* sustainability that, in sociopolitical terms, rests on the ability of the system to claim the loyalty of its adherents. This loyalty, in turn, requires the propagation of a set of values that, in the long run, is acceptable to the populace, and it requires those sociopolitical institutions that make the realization of those values practical.

Required, therefore, are sets of sociopolitical practices that foster cultural as well as environmental sustainability. These need not be entirely separate considerations. Indeed, as suggested above, several prominent streams of environmental thought have argued the existence of a strong link between the sociopolitical and environmental behaviours characteristic of modern industrial society.[7] The point is simply that an ethical affirmation of the value of the world and its inhabitants provides a basis for principles of sociopolitical behaviour that affirm the value of both the human and non-human.

The preceding discussion provides both practical and ethical arguments for sustainability, and some indication of the dimensions of sustainability that are of interest. We will now attempt to interpret those arguments in terms of a definition of sustainability.

Sustainability Defined

A General Definition

In our view, the Brundtland Commission's definition of sustainable development provides an appropriate starting point. Its emphasis on meeting human needs and on sustaining the capability to continue meeting such needs provides a useful focus for applying the concept of sustainability both biophysically and sociopolitically. Another important characteristic of the WCED analysis is the explicit link made between social and environmental problems. At the same time, we need to be wary of treating the symptoms rather than the disease, and of confusing growth with development (Daly 1991; Daly and Cobb 1994), and we need to press for more far-reaching changes in beliefs, attitudes, and social practices than are implied in some of the Brundtland language.

For our purposes, therefore, *sustainability is defined as the persistence over an apparently indefinite future of certain necessary and desired characteristics of the sociopolitical system and its natural environment.*

This definition has several important components. First, the phrase 'apparently indefinite future' has been chosen to reflect the facts that we cannot, and would not want to, guarantee persistence in perpetuity. Such guarantees would be meaningless in practice, because we simply cannot usefully predict that far ahead. Moreover, they would be undesirable in principle, because we don't want to preclude the possibility of desirable change, even in those characteristics of the system we now like. (For example, what we consider desirable and indeed necessary in Canada in 1993 is much different than what the inhabitants of this land, or indeed of Europe, would have said 500 years ago.)

We instead want to ensure, as far as possible, the persistence of those characteristics of Canadian society and its natural environment that we consider necessary and desirable. At the same time, we want to allow for alterations in those characteristics, and in Canadians' interpretations of them. That is, we want to preserve the capacity for the system and the environment to change.

Second, the definition incorporates both necessary and desirable characteristics of the system and the environment. There are several reasons for this. Perhaps most importantly, it is hard to separate necessary characteristics from desirable ones. In the social realm, there is no consensus on the defining characteristics of human society, let alone the necessary conditions of survival for that society, beyond such basics as meeting the survival needs of humans and providing for the nurturance and socialization of the young (Giddens 1984). In the biophysical realm, while minimum conditions of biological survival at the individual level seem straightforward, we do not know enough to describe in detail all of the biophysical processes that maintain those conditions, let alone the necessary states of such processes.

Moreover, the value-ladenness of science and the degree to which it is socially contingent (Clark and Fujimura 1992; Pickering 1992) suggest that it may not ever be possible to determine the necessary conditions of either environmental or societal sustainability unambiguously, or to separate these from merely desirable conditions. These reasons support the view that sustainability is ultimately not a scientific concept but a normative principle.

Third, conceiving of sustainability in normative terms involving necessary and desired characteristics raises the question of who is to decide what those characteristics are and how such decisions are to be reached. This question underlines the importance of the sociopolitical dimensions of sustainability. For sustainability to be acted on, it must be possible for the desirability of environmental and sociopolitical characteristics of the world to be determined and expressed in the political realm. Yet as noted above, the idea of cultural sustainability implies the need for cultural values, and sociopolitical practices and institutions, that are acceptable to the populace. This general acceptance rules out environmental autocracy as a feasible or desirable response to the problems of unsustainability.

Fourth, our definition implies that there is no single version of a sustainable society.[8] We make no claims to be describing the only possible sustainable society for Canada, or the only one that would correspond to the definition.

Fifth, the implications of the concept of sustainability adopted here mean that the term 'sustainable society' is more appropriate for our purposes than 'sustainable development.'[9] As noted above, we do not want to be constrained by traditional views of economic development. Moreover, society is a much broader concept than development, and allows us to encompass some of the wider issues discussed in this chapter. However, we also recognize the need to provide analysis that can connect with real-world concerns, decision-making processes, and the time required for completion of fundamental changes.[10]

Thus, our strategy is to combine an analytical focus on concrete possibilities for sustainability in the relatively near term (i.e., forty years) with explicit consideration of more far-reaching changes beyond then. The scenarios developed in this project do not describe the full set of transformations required in Canadian society. Not only will the transition be incomplete at the end of our story line, but the definition of necessary and desirable characteristics of sustainability can also be expected to change over that time period. The scenarios do describe the characteristics of a more sustainable society at the scenario endpoint and do attempt to outline some of the changes needed if Canada is to move in the direction of such a society. Our scenarios cannot show the changes required to attain sustainability, only some changes that might make Canada more sustainable.

The concept of sustainability outlined here has been described in very general terms. To make it more concrete and useful for analysis, we need to clarify some important definitional problems. The first has to do with the need to describe clearly just what is to be sustained. The second is concerned with the time period over which the sustainable society must be determined. The third has to do with the need to measure or assess sustainability for it to be a meaningful criterion.

What Is to Be Sustained?

Precisely what must be sustained if Canadian society as a whole is to become sustainable? This is not a simple question. First, the definition's emphasis on 'characteristics' of the system and its environment is intended to suggest that increasing the sustainability of a system is not equivalent to maintaining that system in its current form. Such attempts may indeed increase unsustainability. The social and political structure of France in 1785, for example, was not sustainable and could not be maintained in the face of the social pressures that led to the French Revolution. In fact, attempts to preserve unsustainable social practices can contribute to the eventual overthrow or reform of those practices. Similarly, in the ecological realm, the work of Holling (1986) and others has shown that attempts to manage or control certain aspects of, for example, a forest ecosystem can make it vulnerable to collapse as small changes accumulate or low-frequency but extreme natural events occur.

To use Holling's terminology, the goal is not to increase the reliability (resistance to breakdown) of the systems being considered but to increase their capacity to recover from disturbance. The metaphor is that of safe-fail systems (which can fail 'gracefully' without catastrophic repercussions), rather than fail-safe systems (which are less liable to break down initially but collapse entirely when breakdowns occur).

Second, it is not always clear what should be sustained. For example, the changes in European society, culture, and institutions over the past several hundred years have been substantial. Few, if any, characteristics of any European country have been retained untouched, including even the physical boundaries of many countries. Yet we clearly would want to claim that there has been continuity, that it makes sense to talk about 'English' or 'Italian' history or culture. Such considerations suggest the need to be very clear about what we are interested in preserving in the social realm.

In the natural realm, if we want sustainable forests, we must clarify what that means. Are we trying to sustain the current forest community, at its current successional stage? Are we trying to sustain a certain level or type of biomass production, or a certain annual allowable cut? Do we assume that current forest uses will continue to be predominant? Perhaps we should simply try to sustain the capacity to grow forests. This suggests shifting the focus to the soils on which the forests grow. But what about the wildlife in the forests?

One approach to these problems is to recognize that different levels of systems exist, and that sustainability is a property of the highest, most general level being considered, not of the constituent parts of that system. Given this view, it is first necessary to identify the overall system of interest, and then to consider the sustainability of that system as a whole. Using our two examples, we are concerned with the sustainability of a certain type of society and of a particular forest ecosystem. This may or may not mean that any specific components of that society or that forest should be sustained. Rather, it is Canadian society, and the natural processes on which that society is dependent, that should be sustainable. This may or may not require sustainability for particular component subsystems.

This position clarifies some of the issues but leaves important questions unresolved. How do we define Canadian society and the natural processes on which it is dependent? How do we resolve conflicts or trade-offs between different subsystems? And it raises some new problems. Does it make sense to talk about a sustainable society in Canada if the rest of the world is not sustainable?[11]

A third problem relates to the undesirability of certain kinds of sustainability. Clearly, there are some social conditions and practices that we don't want to sustain, such as widespread poverty or crime. Are these to be taken as examples of conditions that are undesirable whether or not they are sustainable, or are they themselves evidence of unsustainability? Some of the longest-lived human societies have been those in which social practices such as slavery, which we would clearly want to label as undesirable, have been prevalent. This suggests that sustainability defined as mere persistence over time is not a sufficient basis for the evaluation of desirability, at least not without being defined in a way that incorporates such concerns.

These examples suggest that important judgments be made about what specifically should be sustained in order to ensure the sustainability of Canadian society. It is also clear that our concern should be more with basic natural and social processes than with the particular forms that those processes take at any time.

The Time Period

The question of the time period over which sustainability is to be assessed is greatly complicated by the existence of an array of social and natural processes with much different time scales. Individual human beings live up to about 100 years; individual trees, depending on the species, several times that long. Some social practices, such as clipper shipbuilding, have quite short lifetimes; others, such as marriage, span millennia of human history. The same is true of natural processes. Moreover, the lifetime of individual members of a class of social practices or biological species is much different than the lifetime of that class as a whole. The lifetime of democratically elected governments, or individual trout, is measured in years, yet democracy and trout are themselves much longer lived.

It is often asserted that the mismatch between the longer time scales of many natural processes and the shorter ones typical of human decision-making is a major reason for the lack of environmentally sound decision-making (Holling 1986). Yet this point needs to be stated carefully. The distinction is not between the long time horizons of natural processes and the short ones characteristic of social processes. As suggested above, many social processes and practices also have long lifetimes.[12] The concern is with the disparity between the time horizon of individual human decisions and that of the natural processes affected by these decisions. However, the same point could be made about social processes. Individual human decisions, often with very short time horizons, may alter the nature and viability of social practices and institutions characterized by long lifetimes and far-reaching effects on human activity. We therefore need to be concerned with the sustainability of both natural and social systems and processes over the longer term.

To a certain extent, the problem of time horizons has been determined for this project by the selection of a forty-year scenario-analysis period. This is longer than typical social decision-making horizons, but shorter than the time scale of many of the natural and social processes with which we are concerned. Therefore, as suggested above, the goal must be to devise scenarios that increase the sustainability of processes over the forty-year time horizon, but that are also directed toward a continued increase in, or at least a maintenance of, sustainability beyond this period. This approach suggests the desirability of also considering the direction of change with respect to sustainability in the endpoint year, and the continuation of that trend beyond 2031.[13]

The Measurement Problem

Space and time variables complicate measurement. If sustainability is a moving and ambiguous target, and if it necessarily manifests itself over periods longer than that of analysis for this project, there would seem to be some serious difficulties in the way of assessing the sustainability of the scenarios.[14] There are several points that need to be made here.

First, as noted above, it is not meaningful to assess the absolute sustainability of any society at any one point in time. The best that is likely, even in principle, is to assess rather crudely the relative sustainability of the society compared with, say, earlier states. And this assessment will depend on developing a rather precise description of what is meant by sustainability and of what is being sustained.

Second, the ability to develop even relative judgments of societal sustainability depends rather directly on the state of knowledge about social and ecological processes in and around that society. There are two dimensions of interest here: the state of knowledge about the nature of those processes (i.e., theoretical knowledge), and the amount of available data describing those processes.

With respect to theoretical knowledge, the basic science of ecological processes is in its infancy, while not only the possibility but also the desirability of formal theory in the social realm is disputed. On the data side, the situation is little better. For example, there is a great dearth of information about even the physical interactions between natural and social systems. Our large national databases and statistical agencies have not been designed to allow the gathering of information on these interactions (Potvin 1989). The result of these factors is that only very approximate types of information are available.

This dismal situation is mitigated on the socioeconomic side by the existence of the Socio-Economic Resource Framework (SERF), the modelling system used in this project. SERF (described in more detail in Chapter 2 and Appendix B) represented not only the most comprehensive set of socioeconomic data in Canada but also a formal simulation capability based on human activities described in the terms most suitable for linkage to environmental processes.

However, as discussed below, the SERF model only represents human activities; it contains no explicit representation of biophysical phenomena. Moreover, those activities are described in terms of physical quantities that are not linked in a clear way with social issues such as justice, democracy, equity, and so on. The upshot is that it was not possible to develop precise measures of sustainability in this project. Instead, only qualitative judgments were possible of whether the scenario outputs from SERF conformed to the scenario design criteria based on the values and principles outlined in this chapter. (See the discussion of scenario design criteria in Chapter 4.)

From the General to the Specific
The problems of deciding exactly what must be sustained in any given case, of choosing an appropriate time period, and of devising useful measures of sustainability suggest the wisdom of retaining our general definition of sustainability. Any attempt to develop a more precise definition would not only run up against these three problems but also run the risk of contradicting our arguments regarding the need to allow perceptions of necessary and desirable conditions of sustainability, and the natural and social systems themselves, to change.

This does not mean, however, that we must resign ourselves to the inability to say anything concrete about sustainability. While the concept itself remains general, its interpretation and application in particular circumstances will necessarily be specific. It will always be possible to develop more concrete principles that apply the general concept of sustainability to the circumstances of interest. In the case of our project, this meant the development of environmental and social principles of sustainability that, if applied, should reduce unsustainability and reinforce sustainability in Canada over the next few decades, as seen from our current understanding and ethical standpoint.

Environmental/Ecological Sustainability
The term 'environment' in the most general sense refers to anything external to some reference point. In systems theory, it refers to anything external to the perceived (or artificially defined) boundaries of some system that can affect the functioning of that system. 'Ecosystems' are populations of plants and animals sharing preferred habitats that interact among themselves (as 'natural communities') and with the abiotic components of their environments. From this ecosystem perspective, the environment includes solar energy, air, water, and minerals, which

all provide 'life support' for living things. The concept of the ecosystem has evolved over the years (McIntosh 1985). It originally placed an emphasis on the importance of nutrient (biogeochemical) cycles and the flows of energy through natural systems. More recently, the spatial patterns of ecosystems have also been emphasized, particularly in the emerging field of landscape ecology (e.g., Moss 1988) and conservation biology (e.g., Scott et al. 1987).

Ecosystems evolve over time, and the general sequences of successional changes have been described in terms of characteristic features and functions (e.g., Odum 1975; Borman and Likens 1979; Holling 1986). Later stages of succession are generally more persistent and self-sustaining than earlier ones, but they are also subject to change through natural agents such as fire, drought, floods, ice storms, tornadoes, and (in the case of aquatic systems) sedimentary in-filling. These changes serve to maintain spatial 'patchiness' in terrestrial ecosystems and concentrated 'centres of biological organization' in aquatic ecosystems. Hierarchy theory suggests that local 'catastrophes' that maintain a mosaic pattern within ecosystems are in fact necessary for overall ecosystem functioning, because they provide space for regrowth and renewal.

Humans depend on ecosystems for basic life support, for natural resources, and for the disposal of wastes. Humans add to the patchiness of ecosystems by transforming landscapes for intensive urban-industrial and agricultural uses. Together with other resource management practices, large proportions of regional ecosystems are in effect maintained as 'anthropogenic subclimaxes,' and often show evidence of degradation in their structural or functional properties (Rapport et al. 1985). Humans also derive important non-economic values from ecosystems, and there is philosophical debate over the ethical or metaphysical responsibilities that humans have for ecosystems. The ethical principles articulated above represent one possible approach to these issues.

To ensure the ecological sustainability of society, the basic goal must be to keep options open and enhance them where possible. Three strategies should thus be pursued.

First, the life-support systems must be protected. In practice, this will require the decontamination of air, water, and soil by assuring the virtual elimination of discharges of toxic substances, especially those that bioaccumulate and biomagnify in organisms. Priority must go to eliminating the release of artificial 'xenobiotic' compounds that are alien to ecosystems not 'preadapted' through evolutionary experience to absorbing and decomposing such compounds benignly. Other kinds of pollutants, such as organic wastes, must be reduced to levels that do not impair ecosystemic functions by overloading their capacities to process them. Environmentally benign farming and forestry practices must be promoted to reduce reliance on pesticides.

Second, biotic diversity must be protected and enhanced, in part through special measures to protect relatively undisturbed or sensitive ecosystems in the overall context of landscape mosaics considered at different levels of scale and detail. Guidance is provided by a hierarchical framework for defining ecological regions in Canada (e.g., Rubec and Wiken 1984), and the diversity itself must be considered from different levels of overall mosaics, natural communities, particular species, and individual populations of species of particular importance.

Third, resource management strategies must maintain or enhance the productivity of ecosystems through careful management of soils and nutrient cycles. Rehabilitative measures will be needed for badly degraded ecosystems resulting from resource extraction, overuse, or pollution.

These three strategies correspond generally to the threefold categorization of environmental issues introduced in the World Conservation Strategy: maintenance of essential life-support systems, enhancement of biotic diversity, and sustainable resource use (IUCN/WWF/UNEP 1980). In addition, strategies must be developed to adjust to climate change, which is expected to occur over the next half century. Even if measures were taken to reduce the accumulation of greenhouse gases causing this change, the trends are practically irreversible over the next couple of decades because of the inertia in many biogeochemical systems, as well as in institutions (International Panel on Climate Change 1992). Some of the projected changes in Canada include (Atmospheric Environment Service 1987):

- a lowering of the Great Lakes by up to one metre
- a rise in coastal sea levels of between 0.2 and 1.4 metres
- shifts in the distribution of forest biomes of up to several hundred kilometres, generally to the north and east
- extensive drier zones and greater variability in crop yields in the southern Prairies
- a northward migration of agriculture in some areas due to longer growing seasons that will, however, be limited by poor soil conditions
- a general increase in the frequency and severity of droughts.

There is considerable uncertainty about the rates at which these changes will occur, and to what extent they will have emerged by 2030. A network of 'early warning' monitoring sites is needed. However, the degree of scientific consensus about the likelihood of such change, and the scale of impacts involved, suggests the need to develop preventive and adaptive strategies. Of particular importance in this regard, given the dominance of greenhouse gas emission associated with energy production and use, is the development and implementation of policies for increasing energy efficiency and switching away from fossil fuels. More generally, the challenge is an institutional one of keeping options open and developing the capacity to adapt or adjust to 'surprises' of all kinds (e.g., Clark and Munn 1986).

Sociopolitical Sustainability

The sociopolitical realm encompasses all human activities and behaviours, including institutional arrangements and activities. For our purposes, it can usefully be conceived as consisting of individual behaviours on the one hand and the activities associated with collective decision-making on the other. Following Dryzek (1987), we will refer to the latter activities under the rubric of 'social choice mechanisms,' or organized approaches for making collective choices. Examples are the market, the legal system, voting, bargaining approaches, rule by command, etc.

Social choice mechanisms interact with individual behaviour in various ways. Some of these mechanisms, such as the market, are based on the aggregation of individual consumption choices, while others, such as rule by command, involve more direct and centralized control of individual behaviours. In practice, the relationship is always mediated by social rules and institutions that translate individual behaviour into collective action and vice versa (Burns and Flam 1987). In every case, however, there exists some distinction, and potential tension, between individual and collective behaviour.

The sociopolitical realm as a whole, therefore, consists of a set of social choice mechanisms, expressed in the form of institutions and social rules, and a set of individual behaviours variously constrained or influenced by those collective mechanisms. One important form that such influence takes is in the set of institutions and rules used within a society for economic and political decision-making.

From the point of view of sustainability, as defined above, two aspects of sociopolitical behaviour are of interest. First, there are the constraints on human behaviour imposed by the need to promote sustainability in the environmental/ecological realm. Second, there are those characteristics of the sociopolitical realm that are desirable in themselves from the point of view of sociopolitical sustainability. We will deal with each of these in turn.

Environmental/Ecological Constraints on Sociopolitical Behaviour

The preceding section of this chapter outlined a set of conditions to be met to increase the sustainability of ecological systems. Most or all of those conditions implied the need for changes in the ways that humans interact with their environment. Such changes will be reflected in different behaviours at the individual and collective level. The issue is the specific nature of these different behaviours, and the degree to which they require significant alterations in, say, institutional design, economic organization, or political decision-making.

One way to assess the sociopolitical requirements for sustainability is to assess the capability of different social choice mechanisms to meet minimal standards of ecological rationality. An important attempt to do this can be found in the work of Dryzek (1987), who argues that none of the social choice mechanisms currently dominant in Western society meets such standards. While there are significant differences in the capabilities of, say, markets or polyarchical political systems to promote ecologically rational behaviour in particular circumstances, they all suffer, he argues, from a common characteristic, which is fatal to any attempt to promote ecologically sound human behaviour at a more general level. This common characteristic is a reliance on instrumental rationality as the basis for human decision-making. Such reliance precludes adequate treatment both of ecological systems characterized by teleology and emergent properties and of questions of ecological or social value that have to do with ends and not means. Dryzek proposes, instead, the development of social choice mechanisms based on the concept of practical reason and radical decentralization of political and economic decision-making.

Dryzek's analysis led him to postulate a set of general criteria that would have to be satisfied in order for a social choice mechanism to be ecologically rational, that is, to have the capacity to resolve ecological problems. These criteria are:

- negative feedback between social choice mechanisms and ecosystems, which permits information about ecological dysfunctioning to be expressed readily and effectively
- coordination across and within collective actions undertaken in response to environmental problems

- robustness or flexibility in response to changing environmental conditions
- resilience in the face of severe environmental disturbance (Dryzek 1987: 46-54).

Dryzek's arguments represent one example of a large literature, noted at the beginning of this chapter, suggesting that the nature of environmental problems requires a fairly significant restructuring of social behaviour and decision-making. His work is cited here because his approach is compatible with ours, whereby general principles of sustainability are articulated in terms of which specific design criteria can be developed.[15] For our purposes, however, it will be useful to translate his rather general criteria into more specific principles, in the context of the discussion above. In particular, we concentrate on the development of several economic and political principles intended to respond to the idea of environmental constraints.

We begin with the issue of growth. There exists considerable disagreement on the question of whether environmental limits to growth in human activities are being approached in practice or whether they can be indefinitely deferred due to technological advances and the resultant substitution of human-made capital for natural capital. But clearly the overall scale of human activity, in terms of both resource use and waste assimilation, must be kept below the total carrying capacity of the planetary biosphere. In this regard, it is useful to keep in mind the distinction between growth and development drawn by Herman Daly (1991:243): 'Much confusion could be avoided if we would agree to use the word "growth" to refer only to the quantitative scale of the physical dimensions of the economy. Qualitative improvement could be labelled "development" ... Growth of the economic organism means larger jaws and a bigger digestive tract. Development means more complete digestion and wiser purposes.' In practice, this principle suggests the need to consider seriously the question of when and how it will become necessary to impose absolute limits on the overall growth in the physical scale of the economy.

In addition to being concerned with the rate of growth of overall activity, we need to consider the substantial negative impact of current economic activity on the natural environment. The issue here has to do with the pervasiveness of what economists call externalities and other forms of market failure connected with the extraction, use, and disposal of environmental resources and amenities. The general principle must be to recognize the environmental cost of these activities, and to incorporate such a recognition in collective and individual be-haviour. In part, this latter goal can be accomplished by minimizing physical throughput per unit of economic activity, that is, increasing efficiency of resource use; improving economy of product design; promoting reduction, reuse, recycling, and recovery of wastes, etc.; and reducing the use of particularly noxious substances or activities. However, it will also be necessary to expend considerable efforts in decontaminating and rehabilitating degraded ecosystems, resources, and amenities. All of these responses will require substantial modifications in economic behaviour. To some degree, these modifications can be accomplished through internalization of environmental externalities via mechanisms such as taxes, effluent charges, and social cost pricing, but there are limits to the ability of such approaches to resolve environmental problems.

The economic implications of the principles outlined here are considerable. And they are not straightforward. As pointed out in the Brundtland report, the most negative environmental impacts are most closely connected to the incidence of both great wealth and great poverty. It does not take much imagination to guess on which end of the economic spectrum the burden of increased environmental concern is likely to fall the most heavily. Moreover, many of the links between environmental constraints and institutional or individual implications are very complex. The connections between, say, changed forestry practices, Canada's balance of trade, employment, economic output, inflation, consumer prices, monetary and fiscal policy, etc., are often indirect and counterintuitive. This means that explicit attention will have to be paid to the equity impacts, as well as the economic efficiency, of the required changes (Gardner and Roseland 1989).

Ensuring such efficiency and equity is essentially a political problem. However, there also exist some more direct political issues raised by the need to implement the environmental/ecological principles outlined in the last section. First, it is necessary, as a matter of principle, to ensure that environmental concerns are incorporated directly into political decision-making in a way not typical of past or current practice. This change is likely to require substantial improvements in environmental assessment procedures, the creation of new legal mandates to provide the power required for implementation of sustainability, and perhaps the development of an Environmental Bill of Rights (Elder and Ross 1989).

Second, that sustainability is a normative ethical principle means that its interpretation in practice is necessarily a political act. This implies a greatly enhanced degree of public involvement in environmental decision-making as sustainability principles, and their application, are debated in the political arena. This is not just a matter of after-the-fact public education and approval. Because the principles themselves are normative, public involvement should occur at the definitional and policy development stages of decision-making.

Third, it is desirable for political activity to be linked more closely to actual environmental experience. This closer link would permit more direct reflection of the causes and consequences of environmental behaviour in political decision-making, allow for the direct expression of individual concern about the environment, and encourage the development and political expression of a relationship to the surrounding environment that can most easily occur at the local level. In practice, this would mean a functional decentralization of some political power to jurisdictions closely linked to natural environmental regions, and the promotion of greater local and regional self-reliance. Such a process has been argued for by proponents of what has been called bioregionalism (Sale 1985).

There are, of course, important limits to such decentralization. Such limits are imposed not only by the need for coordination and consistency of, say, economic activities at the national level but also by environmental imperatives themselves. It will continue to be necessary to develop global, national, and regional responses to many environmental problems. On the other hand, the coordination of political jurisdictions at different spatial levels is a common characteristic of political life. The challenge is to design a set of institutions and rules that permit some functional decentralization of the kind suggested here without compromising the capability of collective responses to environmental concerns at higher political levels.

Sociopolitical Desirability

The preceding discussion provides some principles for collective and individual behaviour compatible with ecological sustainability. By themselves, however, these principles say nothing about the type of sociopolitical system it would be desirable to have, given ecological sustainability. This is important because it would presumably be possible to ensure environmental sustainability in socially undesirable ways (although, as discussed earlier, this would be incompatible with the principle of cultural sustainability). We therefore turn to the question of the sociopolitical realm itself. To do so, we return to the ethical discussion contained in the section on sustainability as a normative ethical principle.

We noted that the self-conscious decision-making and technology-creating capacities of humans have created a cultural evolutionary process that has removed them, to a significant degree, from the determinants of the biological evolutionary process. Such cultural capabilities define the unique character of human potential and the requirements for human expression and fulfilment. They are also the basis of individual moral responsibility. The fulfilment of such potential and the exercise of such responsibility require the opportunity to make meaningful decisions about individual and collective behaviour. Hence, we assert as a principle that the ability of all persons to participate in decision-making about things that affect their lives, the lives of others, and the world around them is a necessary consideration in the design and creation of all sociopolitical structures and institutions.

Meaningful decision-making requires the ability to influence effectively the powers that regulate the interaction of people in a society with each other and with the natural environment around them. Therefore, an open, accessible political process that has effective decision-making power at the level of government closest to the situation and lives of the people affected by a decision is required. Such a process should promote public involvement in decision-making to allow identification and choice of paths of development that are consistent with people's needs, values, and cultural identity.

Responsible participation in decision-making requires freedom from extreme want and from vulnerability to economic coercion, as well as the positive ability to participate creatively and self-directedly in the economic system through which much social interaction and decision-making take place. Moreover, ethical principles suggest the desirability of a minimum level of material equity in society. Thus, all persons should have sufficient wealth and security for themselves and their families to remove them from the possibility of intimidation, exploitation, and coercion of any kind that would inhibit their full participation in political processes.

The ability to engage in good, responsible decision-making requires a minimum level of equality and social justice, including equality of opportunity to realize one's full human potential, adequate material wealth, recourse to an open and just legal system, and freedom from political repression. It also depends on access to high-quality education at all age levels, coupled with effective access to information and information distribution systems. Other important characteristics of the principle of access to information include freedom of religion, speech, and assembly.

A basic assumption of this project was that a sustainable society depends ultimately on a sustainable environment. Thus, the amount of the country's wealth that is distributed as material rewards and incentives is limited on the one side by what is created over and above the amount required to maintain the general populace at a reasonable standard of living and by what can be created without taxing the environment and resource base beyond what they can sustain on the other.

Conclusions

In this chapter, we have attempted to provide a set of principles in terms of which specific design criteria can be developed that will permit the development and evaluation of scenarios of a sustainable society for Canada. We have articulated a set of general principles of sustainability and a set of more detailed, lower level principles for the environmental/ ecological and sociopolitical realms of the project. These principles, summarized in Appendix 3.1 below, are broadly consistent with those found elsewhere in the literature on sustainability issues. When taken together, they define a set of characteristics of a proposed sustainable society for Canada. The next steps in the Sustainable Society Project were to express these principles in the form of scenario design criteria, and to develop and run the scenarios. We turn in the next chapter to a description of how that was done.

Appendix 3.1

Principles of sustainability

Basic value principles
- The continued existence of the natural world is inherently good. The natural world and its component life forms, and its ability to regenerate itself through its own natural evolution, have intrinsic value.
- Cultural sustainability depends on the ability of a society to claim the loyalty of its adherents through the propagation of a set of values that are acceptable to the populace and through the provision of those sociopolitical institutions that make the realization of those values possible.

Definition of sustainability
- Sustainability is the persistence over an apparently indefinite future of certain necessary and desired characteristics of the sociopolitical system and its natural environment.

Key characteristics of sustainability
- Sustainability is a normative ethical principle. It has both necessary and desirable characteristics. There therefore exists no single version of a sustainable system.
- Both environmental/ecological and sociopolitical sustainability are required for a sustainable society.
- We cannot, and do not want to, guarantee persistence of any particular system in perpetuity. We want to preserve the capacity for the system to change. Thus, sustainability is never achieved once and for all, but only approached. It is a process, not a state. It will often be easier to identify unsustainability than sustainability.

Principles of environmental/ecological sustainability
- Life-support systems must be protected. This requires the decontamination of air, water, and soil, and a reduction in waste flows.
- Biotic diversity must be protected and enhanced.
- We must maintain or enhance the integrity of ecosystems through the careful management of soils and nutrient cycles, and we must develop and implement rehabilitative measures for badly degraded ecosystems.
- Preventive and adaptive strategies for responding to the threat of global change are needed.

Principles of sociopolitical sustainability
(1) Derived from environmental/ecological constraints:
- The physical scale of human activity must be kept below the total carrying capacity of the planetary biosphere.
- We must recognize the environmental costs of human activities and develop methods to minimize energy and material use per unit of economic activity, reduce noxious emissions, and permit the decontamination and rehabilitation of degraded ecosystems.
- Sociopolitical and economic equity must be ensured in the transition to a more sustainable society.
- Environmental concerns need to be incorporated more directly and extensively into the political decision-making process through mechanisms such as improved environmental assessment and an Environmental Bill of Rights.

- There is a need for increased public involvement in the development, interpretation, and implementation of sustainability concepts.
- Political activity must be linked more directly to actual environmental experience through the allocation of political power to more environmentally meaningful jurisdictions, and through the promotion of greater local and regional self-reliance.

(2) Derived from sociopolitical criteria:

- A sustainable society requires an open and accessible political process that puts effective decision-making power at the level of government closest to the situation and the lives of the people affected by a decision.
- All persons should have freedom from extreme want and vulnerability to economic coercion, as well as the positive ability to participate creatively and self-directedly in the political and economic system.
- A minimum level of equality and social justice should exist, including equality of opportunity to realize one's full human potential, recourse to an open and just legal system, freedom from political repression, access to high-quality education, effective access to information, and freedom of religion, speech, and assembly.

Notes

1 For excellent annual reviews of the nature and extent of global and regional environmental problems, see Brown et al. (1995); World Resources Institute (1994); and Clark and Munn (1986).

2 Concern over the nature and extent of such global changes led to the establishment in the 1980s of the International Geosphere-Biosphere Programme on Global Change (IGBP), a collaborative international scientific research program (International Council of Scientific Unions 1987; International Geosphere-Biosphere Programme 1992). The IGBP concerns itself with the natural science side of global changes. More recently, the Human Dimensions of Global Environmental Change Programme was established in order to address the human causes and consequences of global change (International Federation of Institutes of Advanced Study 1989; Human Dimensions of Global Environmental Change Programme 1994).

3 There also exist, of course, criticisms of the concept from what might be called an anti-environmentalist persuasion. Authors such as Simon and Kahn (1984), for example, have argued that there are no environmental problems that cannot be solved through continued development of science and technology in much the same way as they have developed in the past. Such writers see the environmental argument as the real threat to society. These concerns will not be addressed here.

4 On the other hand, the concept of sustainable development has the great merit

of forcing a recognition that environmental issues cannot be addressed in isolation from economic and social development issues, especially the problems of poverty, equity, and distributive justice. While we adopt the less loaded term 'sustainability' in this book, we endorse the need to consider economic and social issues, as well as ecological ones.

5 For the purposes of this chapter, the term 'natural world' simply refers to the external biophysical environment within which human beings exist. The line between the natural and the fabricated is notoriously hard to draw, but its precise location is not an important issue here.

6 Of course, the ethical positions outlined earlier have practical sociopolitical consequences in that they constrain the types of sociopolitical behaviour that should be pursued.

7 This is not, of course, to suggest that these schools of thought agree on either the root causes or the appropriate responses to such behaviours, merely that they argue that they are linked.

8 Note that this is a different point than the one made earlier about changes over time. That is, the definition of sustainability provided here implies not only that each definition may be quite different but also that each definition may itself change over time.

9 The term 'sustainable society' was coined by Brown (1981). For a discussion of other uses of the term 'sustainability,' see Brown et al. (1987).

10 Fundamental changes of the type argued for by deep ecologists, for example, are not the sort to be realized in a period of several decades. Students of historical change have noted that the mechanical, Newtonian worldview took several centuries to become dominant in Western society (Butterfield 1957). Ironically, it is precisely this mechanistic worldview that some have described as being at the root of our environmental problems (Berman 1984; Capra 1983).

11 For example, Rees and colleagues have suggested that, because most industrialized countries or regions effectively appropriate the natural capital required to produce the goods and services that they import, the concept of sustainability must necessarily be global. See Rees (1991, 1993).

12 Of course, no social practices have as long lifetimes as those natural processes that change only over geological time. However, such natural processes are relevant to discussions of human behaviour insofar as they can be altered by human activity in the relatively short term.

13 The last year of historical data in the modelling system used in this study is 1981. The scenarios thus start in 1982, and a fifty-year time horizon takes us to the year 2031. Because the scenario analysis was calibrated to real data in 1990, the actual future scenario analysis spanned the forty-one-year period from 1990 to 2031.

14 For a discussion of the measurement problem vis-à-vis the concept of sustainability, see Liverman et al. (1988); Hammond et al. (1995); and Hodge (1996).

15 While Dryzek uses the term 'ecological rationality' rather than 'sustainability,' these terms are not significantly different as they apply to the environmental/ecological realm of this project. A society that has the appropriate social choice mechanisms to meet Dryzek's standard of ecological rationality would correspond to a society that we would consider sustainable in environmental/eco-

logical terms and in terms of environmental/ecological constraints on sociopolitical behaviour. In order to encompass sustainability in the full sense of the term used here, we would add those characteristics of the sociopolitical realm that are desirable in themselves from the point of view of sociopolitical sustainability, which are described in the following subsection.

References

Atmospheric Environment Service. 1987. *Climate Change Digest*. Ottawa: Canadian Climate Impacts Program, Environment Canada, CCD 87-01

Barbier, E. 1987. 'The Concept of Sustainable Economic Development.' *Environmental Conservation* 14:101-10

Berman, M. 1984. *The Reenchantment of the World*. New York: Bantam

Bookchin, M. 1986. *The Modern Crisis*. Philadelphia: New Society Publishers

Borman F., and G. Likens. 1979. *Pattern and Process in a Forested Ecosystem*. New York: Springer-Verlag

Brown, B., M. Hanson, D. Liverman, and R. Meredith. 1987. 'Global Sustainability: Toward Definition.' *Environmental Management* 11:713-19

Brown, L. 1981. *Building a Sustainable Society*. A Worldwatch Institute Book. New York: W.W. Norton

Brown, L., D. Denniston, C. Flavin, H. French, H. Kane, N. Lenssen, M. Renner, D. Roodman, M. Ryan, A. Sachs, L. Starke, P. Weber, and J. Young. 1995. *State of the World 1995*. A Worldwatch Institute Report on Progress toward a Sustainable Society. New York: W.W. Norton

Burns, T., and H. Flam. 1987. *The Shaping of Social Organization*. London: Sage

Butterfield, H. 1957. *The Origins of Modern Science*. Toronto: Clark, Irwin

Capra, F. 1983.*The Turning Point*. Toronto: Bantam

Clark, A., and J. Fujimura, eds. 1992. *The Right Tools for the Right Job: At Work in Twentieth Century Life Science*. Princeton: Princeton University Press

Clark, W., and R. Munn, eds. 1986. *Sustainable Development of the Biosphere*. London: Cambridge University Press

Daly, H. 1991. 'Sustainable Development: From Concept and Theory toward Operational Principles.' *Steady-State Economics*. 2nd ed. Washington, DC: Island Press

Daly, H., and J.B. Cobb, Jr. 1994. *For the Common Good: Redirecting the Economy toward Community, the Environment and a Sustainable Future*. Boston: Beacon Press

Devall, B., and G. Sessions. 1985. *Deep Ecology: Living as if Nature Mattered*. Salt Lake City: G.M. Smith

Dryzek, J. 1987. *Rational Ecology: Environment and Political Ecology*. Oxford: Basil Blackwell

Elder, P., and W. Ross. 1989. 'How to Ensure that Developments Are Environmentally Sustainable.' In *The Legal Challenge of Sustainable Development*. Ed. O. Saunders. Calgary: Canadian Institute of Resources Law

Environment Canada. 1989. 'Conference Statement – The Changing Atmosphere: Implications for Global Security.' Conference sponsored by the Government of Canada, Toronto, 27-30 June 1989

Gardner, J., and M. Roseland. 1989. 'Thinking Globally: The Role of Social Equity in Sustainable Development.' *Alternatives* 16 (3):26-34

Giddens, A. 1984. *The Constitution of Society*. Berkeley: University of California Press

Hammond, A., A. Adriaanse, E. Rodenburg, D. Bryant, and R. Woodward. 1995. *Environmental Indicators: A Systematic Approach to Measuring and Reporting on Environmental Policy Performance in the Context of Sustainable Development*. Washington, DC: World Resources Institute

Hodge, A. 1996. 'A Systematic Approach to Assessing Progress toward Sustainability.' In *Achieving Sustainable Development*. Ed. A. Dale and J. Robinson. Vancouver: UBC Press

Holling, C. 1986. 'The Resilience of Terrestrial Ecosystems: Local Surprise and Global Change.' In *Sustainable Development of the Biosphere*. Ed. W. Clark and R. Munn. London: Cambridge University Press

Holtz, S. 1988. 'Environment/Economy Interactions: Great Expectations.' Presentation to the APCA's Fifth Canadian Governmental Affairs Seminar, Current Governmental Policy and Regulatory Reform – Will It Lead to Sustainable Development? Ottawa, 16-18 October

Human Dimensions of Global Environmental Change Programme. 1994. *Human Dimensions of Global Environmental Change Programme: Work Plan 1994-1995*. Geneva: HDP

International Council of Scientific Unions. 1987. *The International Geosphere Biosphere Program: A Study of Global Change*. Paris: ICSU

International Federation of Institutes of Advanced Study. 1989. 'The Human Dimensions of Global Change: An International Programme on Human Interactions with the Earth.' Report of the Tokyo International Symposium on the Human Response to Global Change, Tokyo, 19-22 September 1988. Toronto: IFIAS

International Geosphere-Biosphere Programme. 1992. *Global Change: Reducing Uncertainties*. Stockholm: IGBP

International Panel on Climate Change (IPCC). 1992. *Climate Change 1992: The Supplementary Report to the IPCC Scientific Assessment*. Cambridge: Cambridge University Press

IUCN/WWF/UNEP. 1980. *World Conservation Strategy: Living Resource Conservation for Sustainable Development*. Gland, Switz.: IUCN

Leiss, W. 1974. *The Domination of Nature*. Boston: Beacon Press

Liverman, D., B. Hanson, B. Brown, and R. Meredith. 1988. 'Global Sustainability: Toward Measurement.' *Environmental Management* 12:133-43

McIntosh, R. 1985. *The Background of Ecology: Concept and Theory*. Cambridge: Cambridge University Press

McLaughlin, A. 1993. *Regarding Nature: Industrialism and Deep Ecology*. New York: State University of New York Press

Moss, M., ed. 1988. *Landscape Ecology and Management*. Proceedings of the First Symposium of the Canadian Society for Landscape Ecology and Management. University of Guelph, May 1987

Naess, A. 1973. 'The Shallow and the Deep, Long Range Ecology Movement.' *Inquiry* 16:95-100

Odum, E. 1975. *Ecology*. 2nd ed. New York: Holt, Rinehart and Winston

Pickering, A., ed. 1992. *Science as Practice and Culture*. Chicago: University of Chicago Press

Potvin, J. 1989. 'Economic-Environmental Accounts: A Conspectus on Current Developments.' Report prepared under contract to the Corporate Policy Group, Environment Canada, by the Rawson Academy of Aquatic Science, Ottawa

Rapport, D., H. Regier, and T. Hutchinson. 1985. 'Ecosystem Behaviour under Stress.' *American Naturalist* 125:617-40

Rees, W.E. 1991. 'Ecological Footprints and Appropriated Carrying Capacity: What Urban Economics Leaves Out.' *Environment and Urbanization* 4 (2): 121-30

—. 1993. 'Pressing Global Limits: Trade as the Appropriation of Carrying Capacity.' Presentation to the Symposium on Growth, Trade and Environmental Values, Westminster Institute for Ethics and Human Values, London, ON, 11-12 February

Rubec, C., and E. Wiken. 1984. *Ecological Land Survey: A Canadian Approach to Landscape Ecology*. Ottawa: Environment Canada

Sale, K. 1985. *Dwellers in the Land: The Bioregional Vision*. San Francisco: Sierra Club Books

Scott, J., B. Csuti, J. Jacobi, and J. Estes. 1987. 'Species Richness: A Geographic Approach to Protecting Future Biological Diversity.' *BioScience* 37:782-8

Simon, J., and H. Kahn. 1984. *The Resourceful Earth: A Response to Global 2000*. New York: Basil Blackwell

Suzuki, D. 1994. *Time to Change: Essays*. Toronto: Stoddart

United Nations Conference on Environment and Development. 1992. *Agenda 21: Programme of Action for Sustainable Development*. New York: United Nations

World Commission on Environment and Development. 1987. *Our Common Future*. Oxford: Oxford University Press

World Resources Institute and International Institute for Environment and Development. 1994. *World Resources 1994/95*. New York: Basic Books

4
Design Criteria for a Sustainable Canadian Society
D. Scott Slocombe, Sally Lerner, and Caroline Van Bers

Introduction

The sustainable society values and principles outlined in Chapter 3 are an essential first step in articulating a vision of a sustainable society for Canada. However, they do not provide much guidance about what such a society might look like, or what changes might be required to achieve such a future. It is therefore necessary to try to flesh out those values and principles in the form of more specific descriptions of what they might mean for both the environment and human activities. These more specific descriptions are called here scenario 'design criteria,' because they were used to guide the development and evaluation of the sustainable society scenario described in Chapter 5. Like the principles themselves, they are divided into ecological and sociopolitical categories.

The purpose of this chapter is to outline specific environmental/ecological and sociopolitical design criteria for a sustainable Canadian society. Seeking sustainability means redesigning Canadian society so that human activities do not have long-term undesirable impacts on the environment or the fabric of society. The design criteria described in this chapter are derived from sustainable society objectives based on (1) the definition and principles of sustainability described in Chapter 3, and (2) constraints that are imposed on the definition and these principles by environmental ecological limits, sociopolitical realities, and culture-specific values. The design criteria represent the attempt to apply these constraints to the definition and principles. The approach was not wholly utopian: neither unlimited funds nor a complete reversibility of problems was assumed. Because these design criteria need to have practical relevance, sectors of human activity (such as forestry and fisheries) were examined in relation to both ecological components (such as forests and fresh waters) and sociopolitical components (such as political and cultural decision-making).

The Sustainable Society Project (SSP) design criteria are an attempt to suggest feasible, medium-length to long-term changes in human activities that will comply with the ecological and sociopolitical constraints involved with achieving societal sustainability. The environmental/ecological criteria derive from an evaluation of the major areas of human interaction with the natural environment in light of our knowledge of ecological systems and resource management. The sociopolitical criteria parallel the environmental/ecological, and are derived from an examination of the sociopolitical realm as one composed of decisions and institutional arrangements.

The design criteria, like the SSP as a whole, suggest how human activity could change in order to implement the ecological and sociopolitical values discussed in Chapter 3. The specific design criteria identified here are not the only ones possible, desirable, or necessary; a complete and certain list is inherently impossible. But they are illustrative of the kinds of changes that are necessary to achieve sustainability. These criteria are one interpretation of how to turn the values and principles identified by the SSP to define a sustainable society into scenarios of how life in such a world might be. The analysis presented here and in the following chapters is just one version of how these principles might be applied, and is intended to stimulate discussion of alternative versions.

Classifying Human Activities

The human activities and environmental/ecological components identified for assignment of SSP design criteria were based on the categories of the stress-response framework (Rapport and Friend 1979) used in state of the environment reporting (SOER) in Canada (Bird and Rapport 1986; Statistics Canada 1986). This is a framework for viewing environmental change as a result of physical alteration, harvesting/extraction, or waste generation. These categories help to identify unsustainable activities and alternatives to them.

More specifically, the components were chosen to represent major societal activity sectors that are mutually exclusive, that conform to available information sources, and that correspond to real economic and ecological sectors. Exclusivity means that, as far as possible, the categories should avoid overlap. And if these criteria are met, then the categories should correspond to those for which data is available, for example, through the federal ministries of Environment Canada or Statistics Canada.

For the realm of human activities, a classification scheme with the following sectors was adopted for use in developing environmental/ecological (E/E) design criteria: urbanization, recreation, agriculture, forestry, fisheries, industrial activities, energy production, transportation, mining, commercial and domestic activities, and technology research and development. Sociopolitical (SP) design criteria for the human activities realm were developed for the same sectors as well as for four additional types of human activity: health care, education, merchandising, and media. Each of these categories is defined in terms of sustainability in the discussion of the design criteria below.

For the ecological realm, a classification scheme for E/E design criteria was adopted that includes the following components: population/settlements, wildlife, agriculture, forests and rangelands, fresh waters, oceans and coasts, atmosphere, lithosphere, and global systems and cycles. For the sociopolitical realm, the classification scheme for the SP design criteria consists of four types of decision-making activities (legal, political, economic, and cultural) and the institutions that implement them.

The choice of sectors and components is meant to represent the full range of major activities and functions found within each of the two realms, as well as to reflect the categories within the computer model used for this project, the Socio-Economic Resource Framework (SERF) (see Chapter 2). Sector and component boundaries are inevitably drawn from an anthropocentric point of view – even the environmental components closely parallel the sectors of human activity. Thus, the subdivisions of the human activities and environmental components may well be comparable, and benefit from a similar organization of available information, yet be open to critical evaluation of how well they reflect the real interconnections of the human-ecological system. Examples of these interconnections include the role of forests, fresh waters, and oceans and coasts in maintaining wildlife and human populations, and the similar ways that agriculture, forestry, fisheries, and mining support urbanization, recreation, and industry.

The approach taken here is appropriate for scenario development. It complements, but does not replace, a more systems-based approach to defining the needs and implications of sustainability that would draw on ecosystem science and social theory to understand how the environment-society system works.

Developing Environmental/Ecological Design Criteria

The development of environmental/ecological design criteria was approached in several steps. These criteria were initially examined with regard to the impact of human activities on each of the three principles for ecological sustainability described in Chapter 3 – life-support systems, biotic diversity, and ecosystem integrity. These were examined in terms of their environmental impacts: waste generation, physical restructuring, and harvesting and extraction, as well as in terms of dysfunctional institutionalized decision-making. The most significant impacts on the natural environment are outlined in a section describing the problems associated with each sector of human activity.

From this information on environmental impacts, a set of objectives providing for environmental health was developed for each of the identified ecological categories used in state of the environment reporting. The goals were also based on the project team's perception of environmental objectives, and on the basis of recommendations in the same sources used to develop the list of sectors and components. The desired characteristics for each environmental component are listed in Appendix 4.1 at the end of this chapter. Here the contents of the table are general targets and actions necessary to forestall the negative consequences of human activities identified in Appendix 4.1. Overlapping objectives between environmental components reflect the limited number of key ecological functions fulfilled by the natural environment.

These objectives were then refined as E/E design criteria for each sector of human activity, and are listed in a subsequent section of this chapter. To supplement the E/E design criteria, the general principles for ecological sustainability as they apply to each sector of human activity were further examined. The basic assumption was that the health and integrity of species and ecosystems must be maintained. This implies that in some areas remedial, as well as protective, actions will have to occur.

Developing Sociopolitical Design Criteria

It is useful to conceptualize the sociopolitical realm of a sustainable society as one composed of choices, and thus as a realm of individual and societal decision-making. For the purposes of this project, it has five components: four broad types of human decision-making *activities* – political, legal, economic, and cultural – together with the *institutions* through which these are carried out.

What constitutes the first three activities is relatively straightforward, although the rules, power relations, and institutions that govern political, legal, and economic activities are highly complex. Less obvious is what is meant by 'cultural' choices and decision-making activities. This is a deliberately omnibus category that permits consideration of those cognitive, emotional, and symbolizing activities not subsumed by the first three categories, but which permeate them, mainly as unexamined assumptions or belief systems, and which play a central role in human society, particularly with respect to social change. These are scripting/sanctioning (or meaning-providing) human activities such as education, religion, literature and the arts, the mass media, and merchandising. Health care can also be included here, although it is more difficult to categorize.

The products of these meaning-providing activities have been characterized as 'theories,' that is, our value-laden beliefs about the way the world is, who we are, what we are doing, and what we should be doing (Schon 1973). For social change to occur and persist, our 'theories' – about ourselves, the world, what is possible and desirable – must change. This cultural component of the SP realm is key to bringing about a more sustainable society, particularly at the community level, where self-conceptions play a central role in the community's sense of mission, empowerment, and efficacy (Hibbard 1989).

Desired characteristics for the year 2030 for each of the five SP-realm components were derived from the SSP principles of sustainability (see Chapter 3). These are descriptions of what the four types of decision-making activities and their related institutions would look like in a more sustainable Canada. Then relatively specific design criteria in the form of guidelines were developed. They can be used to critique and re-design the major sectors of human activity so that sociopolitical choices in and about those sectors take on more of the previously established 'ideal characteristics.'

In Appendix 4.2, the desired characteristics for the five SP-realm components – political, legal, economic, cultural, and institutional – in the year 2030 are suggested. These characteristics represent the SP parallel to Appendix 4.1. On their basis, SP design guidelines have been developed for fifteen major areas of human activity – eleven from analysis of the E/E realm, and four from cultural activity sectors. A brief discussion of the SP-realm aspects of each of these areas of human activity is presented and followed by more specific SP design criteria.

Ecological and Sociopolitical Design Guidelines for Areas of Human Activity

For each sector of human activity presented in this section, a brief overview of the ecological and/or sociopolitical issues and impacts associated with each activity is provided. The design criteria are then listed for the E/E and/or SP realms, and are preceded by a general statement of the sustainability goal for the sector. As discussed earlier, the design criteria themselves are derived from the analysis of the ecological and/or sociopolitical problems associated with each sector, the general principles of sustainability, and the desired characteristics for environmental and sociopolitical components (see Appendixes 4.1 and 4.2 at end of chapter).

At this stage, the problems and solutions in different sectors are considered independently. In Chapter 6, an exploration of the policy implications of a sustainable society reflects the integration of objectives proposed for these individual sectors.

Urbanization

Problem
Urbanization refers to the concentrated conversion of landscapes and habitats to residential, commercial, and industrial buildings, roads, and other types of human installation. The expansion of urban areas in Canada encroaches on highly productive farmlands, forests, wetlands, and other wildlife habitat. Drainage characteristics are altered, and streams are lost. Present urban growth patterns increase housing costs and waste generation, encourage reliance on automobiles, concentrate political and economic authority within the urban area, and lead to a general detachment of people from their natural environment.

Environmental/Ecological Design Criteria
In a sustainable society, the process of urbanization would protect, to the greatest extent possible, natural ecosystems and agricultural land, and would limit the generation of wastes. This approach implies the need for careful and creative planning of residential, commercial, and industrial areas, and ultimately the need for limits on migration to larger urban centres. Additional land conversion should be limited to lands that are of marginal value to agriculture and wildlife. In addition, more efficient use of land, by increasing densities within existing cities, should reduce the demand for these lands.

Design criteria include the following:

- controlling growth and reducing sprawl by increasing urban density, decentralizing activities, limiting expansion to marginal lands, and placing annual quotas on population growth
- managing for urban wildlife by providing more and better green space, and engaging in community forestry
- increasing self-reliance through economic diversification so that production should meet basic needs at household, community, city, and regional levels (e.g., community gardens)
- developing for energy efficiency by building bicycle paths, improving public transit, reducing the need to travel, utilizing better building designs
- improving water quality, and separating sewer and storm drains
- limiting slope, waterfront, and stream development.

Sociopolitical Design Criteria

Possible choices for future physical urban growth patterns include decentralization, intensification in existing urban areas, and implementation of various models of nodal growth. Each has potential advantages (e.g., energy efficiency in intensification models) and disadvantages (e.g., urban sprawl/rural slums resulting from decentralization). The main goal is to curb the environmentally destructive, inequitable, and alienating aspects of much current urban growth. Spatial integration of residence, workplace, school, and daily recreation areas would reduce the need for automobiles and, with the proper design of public spaces, facilitate social interaction.

Design criteria include the following:

- decentralizing political power within large urban agglomerations to promote citizen governance
- soliciting public participation in environmental and social goal-setting exercises for local areas
- developing political and economic incentives to decentralize employment
- locating residential areas near workplaces
- assuring that urban children have frequent opportunities for experience in natural areas
- requiring municipal environmental auditing of all municipal operations
- supporting municipal programs for environmentally sound use of resources

- reducing the role of the market in providing housing by creating land trusts and non-profit housing
- actively promoting higher density that is designed to make optimum use of space.

Recreation

Problem
Recreation refers to the activities that people choose to engage in for a variety of reasons, such as exercise, pleasure, and competition. Many of those activities take place outdoors and far from an individual's home. Some of them have negligible environmental effects; others, such as fishing, motorboating, and snowmobiling, may have significant impacts. In some areas, negative impacts result from the sheer numbers of those engaged in something as seemingly harmless as birdwatching or hiking. Sensitive and valuable resources are disrupted or lost, and wildlife populations may be reduced or displaced. Some activities are energy intensive, and getting to them may require automobile transportation. Conventional definitions of recreation often ignore the generic meaning of the term, and fail to note the recreational value of productive, non-consumptive leisure pursuits such as volunteer work, political participation, and continuing education.

Environmental/Ecological and Sociopolitical Design Criteria
Recreation will be characterized by more varied, low-impact, nature-oriented activities that may be undertaken close to home. Recreation areas, whether in cities or the wilderness, will have to be expanded in order to reduce the ecological impacts of more people wanting such opportunities. Education, in part to increase the acceptability of regulations designed to minimize impacts, will be important. Wilderness recreation will be redirected away from environmentally sensitive areas.
  Design criteria would include the following:
- encouraging contemplative and non-consumptive recreation
- increasing urban recreation opportunities (commercial and natural) to reduce pressure on wilderness and protected areas
- limiting wilderness vehicle access and facilities development
- integrating recreation, agriculture, and forestry activities
- providing the infrastructure to limit pollution disturbance
- banning recreational hunting

- educating for non-consumptive forms of recreation
- encouraging environmental volunteerism and local political involvement as forms of recreation
- promoting benefits of gardening, i.e., self-reliant, healthful recreation that provides fresher, more nutritious food
- protecting natural areas near population centres, and linking them with trails to create green spaces
- encouraging equitable access to controlled ecotourism experiences in interested regions.

Agriculture

Problem

Agriculture includes cultivating and harvesting food and fibre crops, and raising livestock for meat, dairy, and egg production. Conventional agriculture in Canada is an energy- and capital-intensive activity, currently dependent on synthetic chemical inputs to maintain high levels of production. While productivity has increased over the last forty years, the ecosystems that support agriculture have suffered from loss of habitat, degradation, depletion, and pollution. Modern farming methods have reduced biological diversity, and depleted and eroded soils. Persistent chemicals have been introduced into the food chain. Groundwater has been depleted and stream flows reduced. The number of farms has decreased, and smaller farms have suffered from the increasing costs of credit, machinery, and fertilizers. Agribusiness has steadily replaced the family farm, and rural communities are disappearing as a result. Consequently, ownership of the land is concentrated in the hands of fewer people.

Environmental/Ecological Design Criteria

In order for agriculture to be sustainable, it must, above all, operate with respect for the limits of the terrestrial and aquatic ecosystems that support the activity. Ecologically sound agriculture will be promoted through the application of appropriate technology, non-chemical pest-control techniques, and natural fertilizers. The introduction of a greater variety of plant and animal species, and systems of mixed cropping, will help to restore biotic diversity and enhance resilience at all levels. Healthy agro-ecosystems also rely on the use of appropriate land preparation techniques to reduce erosion and protect soil quality. At the same time, priority should be given to protecting prime agricultural land from urban encroachment.

Design criteria include the following:

- improving the efficiency of energy and water use by selecting crops appropriate for locale
- relying more on local inputs, import substitution, and local markets
- reducing the dependence on pesticides and synthetic fertilizers through the use of natural pest and weed control methods and nutrients produced on the farm
- managing animal waste and decreasing the density of animal housing
- reducing erosion through avoidance of cultivation of marginal lands and use of low-impact tillage
- increasing crop rotation and multiple cropping to control pests and maintain soil fertility
- enhancing and protecting genetic resources by maintaining hybrids and wild varieties
- ensuring adequate habitat for wildlife
- decreasing meat consumption.

Sociopolitical Design Criteria

A sustainable system of agriculture and food production will meet demands for high-quality food and fibre without incurring unacceptable economic or environmental costs; in addition, it will distribute income in a way regarded as equitable by the least advantaged participants in the system. Regional self-reliance, closer producer-consumer contact, lower transportation demands, local production adapted to local ecological and social conditions, and an adequate return for farmers would all contribute to a more equitable, and thus sustainable, system of food production.

Design criteria include the following:

- educating people to eat more locally grown foods that are lower on the food chain
- providing incentives and instructions for preparing nutritious, appealing meals with these foods
- requiring environmental audits of large farming operations, and educating small farmers to do audits
- rewarding those who audit by providing subsidies and loans
- encouraging direct sales and more on-farm processing
- discouraging trade that promotes monocropping and depends heavily on long-distance transportation

- promoting farm ownership, and ensuring a fair return to farmers
- mandating full-cost pricing of food and inputs such as irrigation.

Forestry

Problem

Forestry involves the technical and economic management of all forested land. Forests may be managed not only for timber production but also for values such as wildlife, recreation, tourism, and fishing. The most significant application of forestry is tree harvesting. Timber management practices such as clear-cutting and even-aged management (where trees are planted and harvested according to age) have come under increasing public scrutiny because of growing concern over environmental impacts in forest ecosystems. The concern centres on the practices associated with soil degradation and erosion, watercourse sedimentation and contamination, and inadequate replacement of trees, all of which have resulted in poor forest regeneration over large areas. Consequently, large areas of wildlife habitat are lost, and ecological diversity is reduced.

Environmental/Ecological Design Criteria

Sustainable forestry involves protection of forest habitats and a commitment to forest regeneration, so that, among other things, future generations will be able to use forests to meet their needs. Ecologically sound forest management must take into consideration terrain, slope, aspect, and elevation for tree harvesting and regeneration. Timber production would be limited to lands that are best suited for forestry. Other forested lands would be protected as wilderness areas, and they would contribute to community development by providing habitat for hunting and fishing, tourism and recreation.

Design criteria include the following:

- diversifying the economic base of wilderness areas
- reducing industrial forestry by revising and reducing the size of tenures, increasing protected areas, and limiting logging roads
- improving harvesting practices through selective and patch cutting
- limiting log booming in estuaries and streams
- seeking full reforestation, and limiting burning and chemical use
- diversifying second-growth forest species and maintaining old-growth forests
- managing for wildlife and recreation

- reducing the demand for paper and recycling as much as possible
- ensuring the durability of wood construction where used.

Sociopolitical Design Criteria
Sustainable forestry requires citizen involvement in short- and long-term goal-setting for forest management, and requires adequate supervision, monitoring, and accountability to ensure that goals are achieved. To the extent that timber production is an agreed-upon goal, steps must be taken to conserve the resource through reduced demand and adequate reforestation.

Design criteria include the following:
- restructuring taxes to favour the recycling of paper and the use of recycled paper products
- marketing recycled paper vigorously to establish it as a preferred product
- initiating school and community tree-planting projects to foster an appreciation for trees
- requiring the forest industry to develop methods of harvesting and reforestation that ensure the continued viability of forests as a resource
- requiring better environmental accountability from the industry
- promoting the public acceptance of smaller residential units
- encouraging the public to perceive Canadian forests as common property, about which they are entitled to make informed decisions.

Fisheries

Problem
Fisheries, both freshwater and marine, have exceeded sustainable harvest rates in many areas, leading to a steady depletion of fish stocks over the last forty years. The diversity of aquatic life has been reduced, and in some areas exotic species have replaced native species and altered ecosystems. The closure of overexploited fisheries in Atlantic Canada has created widespread economic distress resulting in human suffering, community instability, and political unrest. Similar collapses need to be avoided on Canada's Pacific Coast and other commercial fishing areas inland.

Environmental/Ecological and Sociopolitical Design Criteria
The economic well-being of many coastal communities will depend on maintaining healthy fish stocks. In order for fisheries to recover,

stronger attempts must be made to assess fish populations and to regulate both catch and effort. Such plans must extend to both sport and commercial fishing. Efforts to eliminate exotic species and aid the recovery of native species should be encouraged. Management should be on an ecosystem basis. The careful development of aquaculture and mariculture to meet human needs for protein could reduce dependence on wild-caught fish. In the meantime, communities traditionally dependent on commercial fisheries need government support to diversify and strengthen their economies.

Design criteria include the following:
- controlling commercial and sport fishing by revising commercial licensing, avoiding the introduction of exotic species, managing species on an ecosystem basis, increasing international cooperation, banning driftnet fishing, controlling overfishing, reducing foreign take, and increasing fish consumption to decrease meat consumption
- substituting and managing aquaculture and mariculture
- using fishery by-products for fertilizer
- rehabilitating degraded aquatic and other ecosystems
- providing fishing communities with the incentives and resources to diversify their economies.

Manufacturing

Problem
Manufacturing activities include all secondary production and manufacturing processes that transform raw materials into usable products. These activities have a profound effect on the natural environment in that they use large quantities of natural resources and are the principal source of many persistent pollutants. Industrial activities have resulted in atmospheric pollution, acid rain, eutrophication of waterways, and bioaccumulation of toxic chemicals in the food chain. Workers in the industrial system have often been treated as cogs in the machinery of production, with little regard for their needs for involvement and recognition. Recently, many of these workers have found their jobs taken over by smart machines or moved to low-wage foreign countries where environmental regulations are minimal or lack enforcement.

Environmental/Ecological Design Criteria
Environmentally sustainable industry will involve the widespread use of cleaner technologies and the more efficient use of renewable re-

sources. Toxic by-products and other wastes will be reduced or eliminated by modification of industrial processes, modernization of equipment, substitution of environmentally benign products, and recycling and reuse.

Design criteria include the following:
- reducing greenhouse gases and toxic organics
- reducing the production of solid wastes and water-borne pollutants
- increasing energy efficiency and recycle, reuse, and redesign processes
- dispersing activities, and producing for local needs
- cleaning up past industrial-production and waste disposal sites.

Sociopolitical Design Criteria

Industrial activities should not degrade the environment, mine renewable resources out of existence, or use up essential non-renewable resources before replacements are found. All manufacturing sectors, including the newer information industries, should redesign their operations to involve workers in decision-making about products, processes, and the basic structure of work, including improved access to information about health hazards and environmental impacts.

Design criteria include the following:
- requiring all but the smallest firms to conduct routine environmental and social audits of industrial operations
- developing effective programs to educate management and labour on working together to clean up problems revealed in the yearly audits
- publicizing widely the environmentally 'ten best' and 'ten worst' firms in town (or province) each year (or more frequently), together with the names of their owners and chief executives
- developing programs for management and labour jointly to discuss goal setting in areas such as literacy, child care, and participatory decision-making in the workplace
- actively encouraging worker ownership of firms, with government support contingent on the submission of plans for environmental protection and workplace democracy.

Energy

Problem

The energy sector includes those activities that produce energy, from exploration to development. Our dependence on fossil fuels and hydroelectric power causes significant damage to land and water. Currently, most of the energy comes from non-renewable fossil fuels, and shortages are inevitable. In addition to local air pollution, fossil fuel burning is having widespread effects on the global climate. Transportation of fossil fuels, particularly oil, also poses significant environmental risks. The hazards and risks associated with production, burning, and disposal of nuclear fuels make this source of energy particularly unsustainable at present.

Environmental/Ecological Design Criteria
Priority must be given to energy conservation through reduced consumption and increased efficiency of use. Improvements in energy efficiency are the fastest and most cost-effective means of reducing a wide variety of environmental problems caused by energy production and use. The secondary goal for the energy sector must be to provide safe sources of energy that minimize the impacts of energy production and consumption on the environment. Development of renewable energy resources and a corresponding decrease in dependence on fossil fuels will protect limited natural resources and reduce air pollution.
Design criteria include the following:
- reducing pollution emissions (e.g., carbon dioxide, nitrogen oxides)
- reducing the use of fossil fuels
- reducing consumption, and increasing conservation and efficiency
- supporting the development of renewable and alternative sources by promoting research and development on solar energy, photovoltaics, and efficient, non-polluting vehicles
- reducing energy waste, and encouraging coproduction
- ensuring that adequate waste disposal methods are available
- encouraging end-use appropriateness.

Sociopolitical Design Criteria
There are numerous institutional changes that could bring about improvements in energy efficiency and conservation, and promote the development and use of alternative energy sources. New legislation, policies, and education are the tools that will contribute to more sustainable energy production and use.
Design criteria include the following:
- encouraging governments to work with local utilities to ensure that every building in the community has an energy audit and is properly weather-stripped, insulated, and ventilated for maximum heating efficiency, as well as equipped with water conservation devices and

energy-efficient light bulbs; this efficiency is to be accomplished by using an appropriate billing system, educating the public, and funding people to do the work
- legislating energy-efficiency standards for all household appliances, with tax breaks for lower income purchasers if prices rise
- requiring companies with a certain number of employees to coordinate employee car or van pools
- reducing car use by legislating for bus and carpool lanes, charging more for parking, billing cars for the use of city streets, promoting 'don't drive' days, improving public transit of all types, establishing bicycle paths for commuters as well as for recreation, and promoting the three R's (reduce, reuse, recycle) to decrease car trips made for shopping
- intensifying education through schools and the media on ways and reasons to conserve all types of energy
- creating local involvement in goal and standard setting for community energy conservation by households, schools, churches, business, and industry
- selecting an appropriate and effective mix of regulations, disincentives, and rewards to effect these changes.

Transportation and Communication

Problem
Transportation and communication choices have direct impacts on land through construction of highways, railroads, airports, transmission lines, etc. Transportation technologies also facilitate many other human activities with even greater environmental impacts. Motor vehicles consume large amounts of energy and emit pollutants such as carbon dioxide, nitrogen oxides, and lead, with effects ranging from global warming and acid rain to human health threats. North Americans have structured their lives around the automobile. In Canada, there needs to be a greater effort toward increasing reliance on carpooling, walking, cycling, and public transportation. While there is some current interest in structuring urban centres to minimize commuting, little has been accomplished in this direction.

Environmental/Ecological Design Criteria
A primary objective for transportation is to reduce dependence on the private automobile by improving mass transit systems and redesigning

communities to reduce the distance to work and shopping. All modes of transportation must become less polluting and more fuel efficient. Ultimately, people will travel less as communication technologies replace travel.

Design criteria include the following:
- reducing reliance on the automobile by shifting from cars to buses, from wheels to rails, and reducing highway and airport subsidies
- improving safety and pollution controls, and eliminating the use of fossil fuels
- regulating shipping pollution, and reducing emissions from road vehicles
- using electronic communication to reduce travel.

Sociopolitical Design Criteria
Advances in electronic communication offer alternatives to all forms of physical transportation. There is controversy over whether electronic communication might actually produce more physical travel, and suggestions that the 'electronic cottage' would end up as a depressing electronic sweat shop. But the use of automobiles and airplanes could be substantially reduced not only by measures having to do with energy but also by the imaginative redesigning of many types of work so that electronic communication could replace physical transportation to some extent.

Design criteria include the following:
- removing tax incentives from business travel
- providing incentives for electronic business communication (e.g., via interactive television)
- developing company 'substations,' electronically connected to the head office, where employees in suburbs could work together but still be close to home
- through the media, schools, and community programs, encouraging interest in exploring one's own town and region, as opposed to travelling long distances for vacations.

Mining

Problem
Extraction and production of minerals and metals is an important source of income for some northern communities in Canada. The mining of these non-renewable resources, however, is often environmen-

tally expensive. In addition to overexploiting limited supplies of minerals, the mining industry in Canada impacts large areas of wilderness during exploration and development. Processing ores creates and uses many potentially toxic chemicals that frequently find their way into the environment through tailings and other emissions. Mineral processing also consumes large quantities of energy and water. The mining industry in Canada is concentrated in 'one industry' towns that prosper while mines operate but face problems of survival when a site is abandoned.

Environmental/Ecological Design Criteria
Environmentally sustainable mining operations will employ practices that allow rehabilitation to occur. Priority should be given to protecting fragile plant and animal habitats through effective environmental assessment, monitoring, and rehabilitation. The development of new technologies that minimize environmental disruption, reduce wastes, and improve the efficiency of the extraction process is vital if mining is to be as sustainable as possible.

Design criteria include the following:
- tightening waste disposal and rehabilitation requirements by recycling, recovering, and ending the trade in waste
- increasing efficiency, directing resources to high-quality uses
- encouraging substitution away from mined products
- limiting the use of fresh waters in extraction and transportation
- controlling deep-sea mining and offshore drilling.

Sociopolitical Design Criteria
To be sustainable, the Canadian mining industry must adopt environmentally benign technologies. Not only must the safety and health of workers and their families be a top priority, but workers must also be able to refuse work that is environmentally harmful and to report environmentally harmful company activities. Towns that depend on mining must receive assistance to become more self-reliant through economic diversification.

Design criteria include the following:
- legislating and enforcing requirements and standards for waste disposal, site rehabilitation, use of the best available technology, and mitigation of environmental and social impacts in the area
- providing for adequate shut-down procedures with regard to social welfare

- giving one-industry towns incentives and financial help to protect their natural environments and to diversify their economies
- encouraging mining by fly-in, fly-out crews so that permanent settlements would not develop
- developing more benign substitutes for minerals.

Commercial and Domestic Activities

Problem
Commercial and domestic activities include the private sector production of goods and services that are not industrial in nature, as well as commercial, institutional, and household activities that produce goods and services. These activities have their own direct impacts, but also drive industrial activity and its associated impacts. Excess packaging and consumption result in inefficient resource use. Large amounts of energy and land are consumed to support consumer habits, and significant quantities of toxic and solid wastes are generated.

Environmental/Ecological Design Criteria
The major emphasis in this sector will be on reducing overall throughput of energy and material per unit of useful activity by use of the three *R*'s, improved energy efficiency, and the purchase of environmentally benign products.
    Design criteria include the following:
- reducing throughput and increasing efficiency of use by reducing, recycling, reusing, and recovering; reducing land and energy requirements, and increasing the density of land use; and reducing water use (e.g., drought-resistant lawns)
- eliminating wildlife products
- reducing packaging to improve solid-waste problems
- distributing activities within and between cities.

Sociopolitical Design Criteria
Because domestic activities involve every person in society, and commercial activities are spread across a wide variety of sectors, changing practices in these activities will require both lifestyle and attitudinal changes in the direction of reducing the consumption of environmentally destructive goods and services. The public should be included in decision-making about the needed changes and how to implement them.

Design criteria include the following:
- helping organizations to develop waste management programs, and educating to promote compliance
- legislating the amount of waste materials that will be accepted from households and organizations
- instituting fee schedules that reward 'three-*R* behaviour,' with the emphasis on reduction
- encouraging environmentally friendly products and processes by identifying, rewarding, and subsidizing them
- educating consumers to demand environmentally friendly products
- through schools, the media, and local service and neighbourhood groups, educating and encouraging all age groups to practise the three *R*'s at home and to substitute benign products for environmentally harmful ones
- requiring reduced packaging.

Technology

Problem
Technology includes the research, development, and production of devices and practices that support and facilitate other activities. While technological development, such as biotechnology and weapons research and testing, has impacts of its own, its greatest impact comes from the use of technology in other human activities. Some technologies are more environmentally destructive and socially undesirable than others, and they interact with social processes and institutions in complex ways. As a result, technologies are not neutral tools with respect to environmental and social impacts, but tend to reinforce dominant patterns of development.

Environmental/Ecological Design Criteria
The use of technology has contributed to almost all of the problems associated with human activities. Yet technology also holds the promise of mitigating some of the worst environmental ills and promoting environmental soundness. It is a matter of developing resource-efficient, non-polluting technologies, and recognizing the fact that new technologies do not need to be large and expensive to be effective. The sociopolitical activities of increasing technological literacy to facilitate informed public choices about technology, and an ultimate reorientation of research and development away from the military, will be critical.

Design criteria include the following:
- shifting research and development away from environmentally harmful areas, approaching new technologies (e.g., biotechnology) cautiously, and encouraging conservation and purification technologies
- developing less toxic, biodegradable substances
- increasing public access to, and transfer of, technology.

Sociopolitical Design Criteria
The development of environmentally and socially sustainable technology requires an institutional infrastructure oriented toward such development processes. A more informed public should be centrally involved in debating goals for the allocation of societal resources to technological research and development. As well, publicly funded research and development should provide some tangible public benefits, rather than just profits for private organizations.

Design criteria include the following:
- providing financial incentives for Canadian firms to develop environmentally relevant technologies for monitoring, prevention, and cleanup
- encouraging worker involvement in developing workplace technologies, and in designing worker roles in the use of technology
- educating for public technological literacy, and developing processes for public participation in decisions about developments in fields such as bioengineering
- providing public 'dividends' when publicly financed research and development pay off.

Health Care

Problem
Health care in Canada is among the best in the world. However, preventive-health and well-being programs are still rare, and there are inequities in health risks due to differential spatial distribution of pollutants, unsafe workplace practices, and so on. Inequities in income, employment, and child care also contribute to basic differences in health status between groups.

Sociopolitical Design Criteria
Building on the best features of its existing health care systems, Canada should ensure that all of its citizens have access to adequate well-being

counselling, preventive programs, and health care, and that costs are controlled to make the best use of societal resources. Existing inequities in health risks due to differential spatial distribution of pollutants should be considered in setting priorities for environmental remediation. The basic inequities in society affecting health status should be examined and remedied.

Design criteria include the following:

- ensuring that there is adequate affordable medical care and preventive health education for all
- educating people to take responsibility for their own health, and to demand safe and healthy surroundings in their communities, homes, and workplaces.
- strengthening and enforcing legislation that allows workers to refuse unsafe work and permits class-action suits in pursuit of a safe and healthy environment.

Education

Problem

Education in the formal classroom has traditionally been designed to equip Canadian students with basic skills of literacy and numeracy as well as with attitudes of respect for, and obedience to, authority. In a simpler world with a strong need for assembly line workers and a servant class, these educational goals seemed appropriate to those who shaped schooling in the service of the dominant culture. More recently, emphasis has been placed on drawing students into science and technology, because these fields are seen as the engines that will drive the Canadian economy in the twenty-first century. While environmental education in various forms is increasingly finding a place in the classroom, it is too often seen as an option rather than as an essential component of basic education.

Sociopolitical Design Criteria

Different approaches to education will be required in the future. People need to develop the qualities of an active citizen: the ability to research and think through the merits of competing social and environmental options, to question authority in an informed and constructive way, and to wield authority in order to involve the broadest possible number of informed stakeholders. No single type of 'job skill' guarantees future employment; rather, students need to develop the learning skills that will see them through a lifetime of self-education.

Design criteria include the following:

- ensuring that all students acquire excellent foundational skills in literacy and numeracy
- stressing self-directed learning from the earliest years
- educating at all levels for the individual's effective participation in social and environmental goal setting, beginning with parent-child discussions and continuing in the schools as children participate in deciding on goals and standards for their classrooms and school
- promoting cooperation rather than competition between students
- promoting mutual respect by soliciting contributions from every student in order to achieve common goals.

Merchandising

Problem
Merchandising drives the Canadian economy. Ours is a buying and selling culture in which people are socialized to think of themselves as consumers, and to consider this role of prime importance in creating meaning and happiness in their lives. The result is a strong cultural emphasis on the consumption of material goods and services, with resultant environmental (and social) implications.

Sociopolitical Design Criteria
New approaches to merchandising are needed. If the volume of energy use and of materials put through manufacturing and other processes is to be reduced, new standards for judging goods and services, such as quality, durability, and environmental appropriateness, must be promoted.

Design criteria include the following:

- using social marketing to promote the virtues of the three *R*'s, particularly the reduction of consumption
- educating the public about the pleasures of non-consumptive activities
- replacing convenience and disposability with durability and reparability
- mandating full-cost pricing and full-disclosure labelling
- developing valid, adequate standards for identifying socially and environmentally benign products

- utilizing social marketing – especially environmental success stories – to reach the public with new sustainability ideas, programs, and opportunities
- encouraging parents to monitor and control the amount and type of television that their children watch, and also to provide feedback to advertisers.

Media

Problem
The mass media, particularly television, absorbs a great deal of time for a majority of Canadians. The mass media sell consumerism and a wide variety of goods, as well as role models, lifestyles, and values.

Sociopolitical Design Criteria
If social change toward greater sustainability is to occur, it would be useful, if not essential, to enlist the commercial as well as the non-commercial media in promoting that change. Television, for example, can be a powerful tool for education and social change.
    Design criteria include the following:
- encouraging a move toward requiring local cable television stations to provide local groups with adequate instruction and other support so they can create and broadcast their own programs about local issues
- encouraging television sponsors, producers, writers, and actors to incorporate environmental and social equity themes into prime-time shows
- educating media consumers to evaluate media content and advertising with social and environmental goals in mind, and to relay opinions to media producers
- providing adequate resources for public service media so that they can compete with commercial media.

Conclusions
No list of design criteria for a sustainable society can be definitive, and those presented here are just one possible set. Not all of these criteria, or even most of them, were incorporated quantitatively in the SERF scenario analysis; some of them were, while the rest formed the context within which decisions about SERF input variables were made. It was not possible to express all of these criteria in terms of the model used in the study; moreover, as we moved from the goal of sustainability through to design criteria for a future society and then to scenarios, the

scope for disagreement regarding the concrete implications of agreed-on general principles widened.

All the same, the design criteria tables are intended to be suggestive, heuristic, and substantially acceptable. Focus on a smaller geographic area, perhaps a region, is likely to increase agreement on design criteria through persons having more common backgrounds, concerns, and goals (Aberley 1985; Staples et al. 1988). Looking at efforts in other regions may also shed light on topics such as public or private sector roles, interregional effects, transferability, and phenomena deserving of further investigation (Chapman 1982).

Society is often trying to avoid unsustainability in a very immediate way. Trying to avoid catastrophes suggests possible opportunities for learning. For example, outlines of general societal feedback loops that affect sustainability should probably be included in scenario design. Design criteria tables could be useful in identifying areas where such feedback is necessary and possible: for example, between developers of urban areas and associated transportation facilities and those living in, planning, and managing the social and environmental dimensions of communities.

The next chapter presents the scenario of what a sustainable society in the year 2030 might look like, based on these design criteria, the sustainability principles, and the results of the SERF computer modelling.

Appendix 4.1

Desired environmental goals in a sustainable society

Population/settlements
- people usefully engaged (vocation and avocation)
- people adequately fed and healthy
- zero population growth
- resource consumption and waste production minimized
- land, air, and water impacts minimized.

Wildlife
- inventory, conserve, and monitor the diversity
- devise networks of protected areas for conservation
- end the trade in luxury wildlife products
- ban the use of wildlife in product testing
- control wildlife harvesting (subsistence only), and protect endangered species.

Agricultural lands
- reduce energy and chemical intensity
- remove degrading (e.g., saline) lands from production
- avoid using steep slopes and marginal lands
- maintain hybrids and wild varieties; increase crop diversity
- increase organic, traditional farming methods; use crop rotation
- use local inputs, summer fallowing, low-impact tillage, and agroforestry.

Forests and rangelands
- preserve the functional role of these lands within the biosphere
- curb acid rain and air pollution
- increase the flexibility of forest- and range-tenure systems
- avoid clear-cutting where it will be damaging
- increase reforestation, rotational grazing, and forestry
- tighten forest and rangeland stumpage and rent systems
- manage for wildlife, recreation, and diversity
- promote non-forestry economic development for the wilderness.

Fresh waters
- improve efficiency in irrigation, domestic, and industrial use
- limit megaprojects and diversions: i.e., no more large dams
- maintain low demand; create high levels of recycling and reuse
- reduce overfishing, and approach aquaculture with caution
- control sport fishing and the introduction of sport species.

Oceans and coasts
- reduce sport and pelagic overfishing
- protect and manage coasts and estuaries
- control mariculture and hatchery developments
- reduce pollution and improve remediation
- protect their role as sink and source
- implement the 1982 Law of the Sea Convention
- develop ecosystem-based fishing regulations; monitor and enforce them.

Atmosphere
- curb pollutant emissions at all levels
- ban CFCs, recapture coolants, and use substitutes
- maintain wind and sun access rights
- stop using fossil fuels; develop solar and wind energy
- maintain the international status of the atmosphere.

Lithosphere
- maintain cover by planting trees and using terracing
- control the population, and avoid the development of marginal lands
- redistribute land, and reform land-tenure systems
- control the development of non-renewable resources
- reduce toxic inputs to the lithosphere.

Global systems and cycles
- reduce regional vegetation and climatic changes
- reduce greenhouse gas (carbon dioxide, ozone, etc.) emissions and production; stop using fossil fuels
- reduce toxic contamination
- plant more trees; stop deforestation.

Appendix 4.2

Desired sociopolitical goals in a sustainable society

Political decision-making in 2030
- reallocate (decentralize) some democratically selected powers from federal and provincial levels to regions and municipalities
- provide mechanisms to ensure that there is as much public participation as possible in setting environmental and social goals, as well as in debating the trade-offs (e.g., economic) involved in reaching them
- develop new social contract and social control mechanisms to ensure that environmental standards and sociocultural rights are maintained
- ensure equity, along with dignity and security, by developing an integrated, more comprehensive social welfare program with a guaranteed annual income at its centre
- ensure equal and broad access to high-quality, lifelong education
- mandate (and fund) research and full disclosure regarding the environmental impacts and dollar costs of all human activities (for input into full-cost pricing)
- require that politicians explicitly take on stewardship duties with full accountability
- safeguard the rights of minorities, as well as the basic civil rights of all.

Legal decision-making in 2030
- create an Environmental Bill of Rights based on principles of sustainability

- create a process for broad public participation in environmental and social target setting
- create a more effective and efficient environmental assessment process that would come into play at the goal and target-setting levels to ensure better evaluation of the environmental costs and benefits of proposed policies, programs, and projects in both the public and private sectors
- create legal mechanisms to ensure integrated, comprehensive research, monitoring, and reporting on the state of the Canadian environment that would function effectively under the conditions of decreased central control
- create legal mechanisms to ensure compliance with standards of environmental quality, minority group rights, and basic civil liberties under the conditions of decreased central control.

Economic decision-making in 2030
- foster economic self-reliance to the extent feasible, from the national to the community level, with each level doing only what the level below cannot do
- devise and implement least-cost strategies, i.e., act to become economically and physically more efficient in the use of resources to meet environmental objectives and generate wealth, thus reducing Canadian resource use and permitting an increase in Third World living standards within global environmental constraints
- constrain the role of market mechanisms by (1) setting limits (based on basic environmental constraints and politically determined social goals) beyond which they may not operate, and (2) affecting prices (through taxes and charges on goods, services, and resources) so as to reach politically determined standards and targets within the basic environmental constraints
- institute full-cost pricing, based on cradle-to-grave responsibility, of all goods, services, and resources
- create appropriate safeguards for those with lower incomes who may be disadvantaged by pricing policies
- institute either a wealth or inheritance tax and eliminate loopholes in the existing tax structure in order to generate revenue needed to maintain social welfare through the transition to full-cost pricing.

Cultural decision-making in 2030
- encourage education, religion, and other cultural media to emphasize

the right of every person to the basic necessities of life, including clean air and water, adequate food and shelter, and health care
- assure access to lifelong learning and meaningful employment for all
- incorporate environmental concepts and materials in all types of formal education and in all subject matter in the curriculum
- give high priority to assuring that children have access to outdoor experiences in natural areas of various types, and that at minimum these experiences instil an affection for, and sense of identity with, the natural world
- promote interest in, understanding of, and protective concern for the ecosystem(s) of one's own region
- reduce dependence on the purchase and possession of material goods as a source of self-esteem, status, and entertainment
- promote the idea of personal and organizational responsibility for preventing environmental degradation and social inequity
- provide frequent participation for all age groups in a variety of group strategies for reaching agreement on social and environmental goals and standards
- create a comprehensive environmental information-gathering system that assures timely feedback and the accurate targetting of information receivers
- provide training and assistance for monitoring and reporting environmental conditions at every level of society, beginning with volunteers in their own communities and workplaces
- encourage networking between decentralized units such as municipalities to secure better diffusion of innovation and better feedback about what works and what doesn't
- promote a norm of cooperation to secure better sharing of information and thus (1) more and higher quality information through inputs from actors with different perspectives; (2) more effective planning for the sustainable use of resources, i.e., for preventing tragedies of the commons; and (3) more efficient uses of resources through planned cooperative allocation.

Institutions in 2030
- contribute to meeting basic human needs
- refrain from activities that degrade the natural environment.
- adopt the most environmentally benign technologies and practices available in the most timely manner possible

- change reward structures to promote more sustainable choices and behaviours
- provide independent watchdog agencies with sufficient and timely information to allow monitoring in terms of the preceding guidelines
- enhance self-reliance at the most local level feasible by employing local people and resources and by educating all stakeholders
- develop a consultation process to involve all informed stakeholders in goal setting (policy), program/project planning, and the evaluation of outcomes
- develop coordinated information-gathering and action-forcing capabilities to provide timely, relevant feedback and appropriate action with respect to the effects of institutional decisions/activities on humans and the natural environment
- create an influential, non-partisan (civil service) Office of Future Scanning to consider long-term institutional challenges and act as a counterweight to short-term, bottom-line thinking
- maximize the versatility of roles and role incumbents, which are key to maintaining either robustness (the ability to function well in a variety of situations, including new ones) or flexibility (the ability to function well in frequently changing conditions).

References

Aberley, D.C. 1985. *Bioregionalism: A Territorial Approach to Governance and Development of Northwest British Columbia.* MA thesis, SCARP, University of British Columbia

Bird, P.M., and D.J. Rapport. 1986. *State of the Environment Report for Canada.* Ottawa: Supply and Services, Canada

Chapman, M.P. 1982. 'The Mature Region: Building a Practical Model for the Transition to the Sustainable Society.' *Technological Forecasting and Social Change* 22:167-82

Hibbard, M. 1989. 'Issues and Options for the Other Oregon.' *Community Development Journal* 24 (2):145-52

Rapport, D.J., and A.M. Friend. 1979. 'Toward a Comprehensive Framework for Environmental Statistics: A Stress-Response Approach.' Statistics Canada Occasional Paper 11-510. Ottawa: Statistics Canada

Schon, D.A. 1973. *Beyond the Stable State.* Harmondsworth, Eng.: Penguin

Staples, L., D. McArthur, J. Butler, B. Green, T. Penikett, P. Boothroyd, M. Decter, J. Potvin, and J. Simard. 1988. 'Mining the Popular Wisdom: Yukon 2000 Brings a Fresh Approach to Northern Economic Development.' *Northern Perspectives* 16 (2)

Statistics Canada. Structural Analysis Division. 1986. *Human Activity and the Environment: A Statistical Compendium.* Ottawa: Supply and Services, Canada

5
Life in 2030:
The Sustainability Scenario
David Biggs and John B. Robinson

Introduction

This chapter is divided into two parts. Part A describes, on a sector-by-sector basis, (1) the scenario input assumptions made about how Canada would need to evolve over the period from 1990 to 2030 in order to conform to the sustainability principles and design guidelines developed in the project and outlined in Chapters 2 and 4, respectively, and (2) the results of the scenario that emerged from running these input assumptions through the SERF modelling framework, and iterating them as required to ensure internal consistency and conformity with the sustainability principles and project design criteria outlined in earlier chapters. Rather than outlining the input assumptions for each sector first, followed by the output assumptions for each sector, we have chosen to present them together, and to provide an integrated picture of each part of Canadian society. Part B describes the institutional changes in legal and political decision-making processes that we assumed for the scenarios, and some of the general environmental conditions that might be expected to result from our scenario.

Part A: Model Inputs and Results

The sectoral breakdown used in this project was derived from the conceptual hierarchy of the SERF system. There are forty-three individual submodels (calculators) in SERF, but for convenience in making qualitative descriptions, they were aggregated into the following twenty sectors, organized into three categories:

Demography: population, labour and the nature of work.

Consumption of goods and services: housing and urban design, consumer goods and services, health care, education, media and communications, food, recreation, transportation, government, office sector, retail trade, infrastructure.

Manufacturing and resources: manufacturing, agriculture, fisheries, forestry, mining, energy production.

Based on SERF's calculations and the qualitative design criteria, an integrated scenario of a sustainable society in 2030 was produced. These results are not predictions, and no effort has been made to assess the likelihood of this scenario. Rather, the description that follows represents one alternative that might result if Canada made a determined effort to become as sustainable as possible using the most environmentally benign technology either available or under development in 1990. As discussed in Chapter 3, the scenario focuses on technological, behavioural, and institutional options, rather than on economic issues. In each case, the discussion of input assumptions and results is written from the viewpoint of 2030, describing the main changes that have occurred over the preceding forty years.

This chapter also describes some of the institutional dimensions of our sustainability scenario, and the environmental conditions that this scenario might be expected to create. This description of possible political, legal, cultural, and environmental changes was developed in advance of the quantitative scenario analysis, and used to guide the development of the scenario inputs; it was also used as a basis to assess the results. It is not tied in any quantitative sense to the formal modelling analysis, but represents a picture of conditions judged to be consistent with those resulting from the SERF analysis.

Demography

Population
The Canadian population stabilized at thirty million in 2000, and has remained at that level. Since 1990, fertility rates have held constant at an average value of 1.67. Divorce rates have declined slightly from their late twentieth-century high, but the numbers of non-traditional family groups have continued to grow. While some demographers speculated that changes in the traditional nuclear family structure would have significant effects on population growth, the results in 2030 have not borne out this view. Instead, Canadians seem to have adopted more diverse family structures without significantly changing childbearing behaviour, perhaps because the security and flexibility provided by the

guaranteed annual income, strong community ties, and high labour force participation rates among women enables childrearing in a variety of family arrangements.

Emigration has held constant at the average levels of the late 1980s, while the immigration rate has reflected a policy decision at the national level to move toward a stable population of thirty million. In fact, immigration has been treated in SERF as a residual; that is, immigration rates have been set to cause the total Canadian population to stabilize at about thirty million by 2030, partly in order to see how sustainable Canadian society appears to be under conditions of population stability. Immigration levels have recently increased somewhat because much of the population is past childbearing years. Mortality rates have also remained nearly constant since 1990, with death rates for those below eighty decreasing slightly and average maximum life span staying constant (80.8 years for women, 73.7 years for men). The result is a slight increase in average life span.

Some interesting trends appear as the future population scenarios are examined. The large number of Canadians under the age of forty-five in 1990 represented a 'baby boom bulge,' which moves through the scenario time period with the result that the percentage of people over sixty-five has doubled from 11 per cent in 1990 to 22 per cent in 2030, as shown in Figure 5.1. The implications of this aging population trend are manifest in several sectors of the economy.

Labour and the Nature of Work
During the late twentieth and early twenty-first centuries, labour force participation rates have remained constant, reflecting lower participation for men and higher for women, who achieved pay equity in virtually all jobs by 2000. The growth in women's participation rates and the decline in men's resulted in both men's and women's rates stabilizing at 80 per cent early in the twenty-first century.

Both men and women in their main childrearing years (late twenties and early thirties) have significantly lower labour force participation rates, ranging from 70 to 75 per cent. This range largely reflects a policy decision made in the early twenty-first century to pay one parent a child-care wage for the first three years of a child's life. Doing so has not only increased the number of families with one parent staying at home but also saved money relative to the state-supported child-care infrastructure, which would otherwise have been required for that child. Of course, institutionalized child care continues to be the preference of those parents who both want to work outside the home.

Figure 5.1

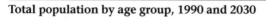

Total population by age group, 1990 and 2030

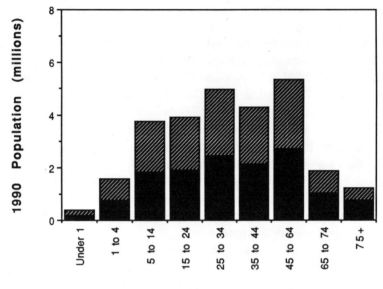

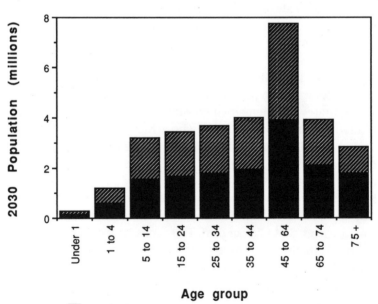

While labour participation rates have remained high, the average work week in the formal economy is 27.5 hours. Canadians increasingly devote some of the working hours of the week to unpaid activities in the informal economy, such as community service or child care. The resultant loss of cash income is partly offset by the higher participation rates and by a growth of bartering and payment in kind for many services. In 2030, people continue working past the age of sixty-five, as the mandatory retirement of the previous century was overturned by the Supreme Court of Canada in 2010.

The result of these changes is a labour force in which most people work actively in both the formal and the informal economy. Taxes and product prices are higher because environmental and social costs are now internalized, but this rise is partly offset by increased reliance on the informal economy. Shared work is increasingly common, and a wide range of employers are now accustomed to accommodating employees who share full-time positions.

The factors affecting the size of the labour force, including changing participation rates and the aging population, resulted in an increase in the total labour force from 13.4 million in 1990 to 15.3 million in 2010, and down to 14.5 million in 2030. The share of women in the labour force shifted from 40 to 50 per cent over the scenario period.

The demand for labour is, of course, determined by the level of activity in each of the sectors of the Canadian economy and the relative labour requirements of those sectors. The scenario described in this chapter resulted in nearly constant demand for labour (measured in person hours) throughout the scenario period. When population growth is taken into consideration, however, the labour demand per capita fell by 14 per cent between 1990 and 2030. Note that the activity of those working and bartering in the unpaid and alternative sectors is not reflected in these figures.

Due to the aging population, the unemployment figures drop during the latter part of the scenario period, as shown in Figure 5.2. Full employment (4 per cent unemployment) could be maintained between 1990 and 2030 with only minor fluctuations in the average work week, which varied between 28.5 and 31.3 hours over the scenario period.

Increasingly, people have work, often in small cooperative ventures, that is supportive of sustainability. Environmental restoration, research, communication, and education are popular, as is work in social planning and community consultancy. Competitive companies are making efforts to introduce sabbatical leaves, employee decision-

Figure 5.2

Unemployment rate, 1990-2030

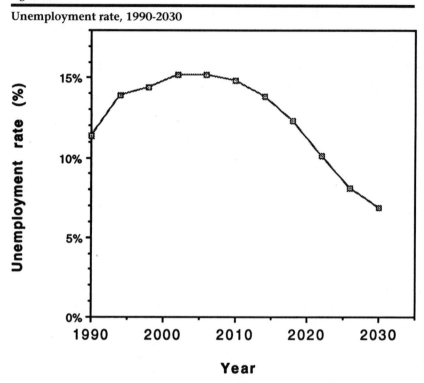

making, sustainability policy statements, and other practices to improve the quality of working life and productivity.

The aim is to ensure a viable, balanced, productive, and self-reliant economy that provides the basic necessities of life for all through a guaranteed annual income (GAI). The GAI was established because the transition to a sustainable society by necessity involves major structural adjustments, and it proved to be more politically palatable and economically efficient to collapse the income-support programs of 1990 into a GAI program. It has also been a more effective way of relieving the cycle of perpetual poverty that many unskilled trades workers and seasonal resource industry workers were vulnerable to in 1990.

Consumption of Goods and Services

Housing and Urban Design
Increased recognition of the environmental and social costs of suburban, commuter-style living (e.g., sewage treatment, road construction,

land use losses, and resource use and disposal), reinforced by the internalization of such costs in the prices of single-family homes and in the cost of travel (e.g., carbon dioxide emissions, road materials, and energy use in construction), have led to an increased emphasis on urban densification. The effects of these developments have been the reclamation of suburban land for new urban nodes and a virtual halt to urban sprawl. No new suburbs are being created around major urban centres. Instead, the conversion of suburban housing tracts to denser cohousing and row housing with their own employment bases, urban amenities, and transit lines is well under way. About 25 per cent of Canadians live in rural areas, and the rest in small urban centres – what used to be the suburbs. These centres are, for the most part, complete communities in terms of employment, services, residences, and so on. The move toward increased cohousing has also strengthened the trend toward sharing home-maintenance equipment and labour, vehicles, and even recreational and cooking space. About 40 per cent of families now live in cohousing units of this kind. Due to the stable population over the last forty years and the durability of buildings, housing starts have plummeted. As a result, changes in the housing mix have been slow (see Figure 5.3).

Figure 5.3

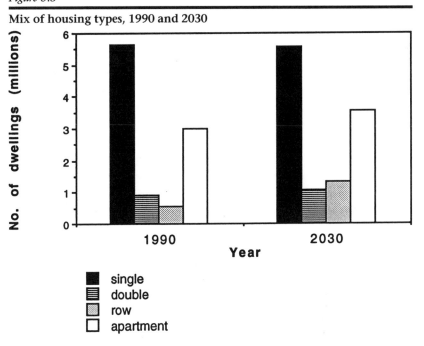

Mix of housing types, 1990 and 2030

The increase in cohousing has resulted in a steady rise in the numbers of dwellings that are inherently more energy and materials efficient. As well, residential landscaping routinely relies on native vegetation, requires virtually no watering or chemicals, and contributes to energy conservation by its positioning relative to each dwelling so as to reduce the house's exposure to both heat and cold. Housing is perceived as a functional good that should require as few resources as possible to construct and maintain. Energy and water efficiency, ease of maintenance, good design, and quality of work are particularly valued.

The gradual introduction of new building materials, many derived from recycling operations, and improved insulation and better heating systems, have produced stronger and quieter structures. Research into indoor air quality had, by the late 1990s, become integrated with heating system design, so that more efficient and environmentally benign heating systems (characterized by improved heat storage, ventilation, and heat recovery) became the norm. In addition to these technical advances in new housing, a massive insulation retrofit program for existing housing has dramatically improved the insulation levels of the entire housing stock, as shown in Figure 5.4. As a result of these changes, energy use per household for home heating has declined on average by 60 per cent, as shown in Figure 5.5. There has also occurred a significant switch from oil and electric baseboard heating to electric heatpumps, while natural gas systems have continued their overall dominance of the heating market (see Figure 5.6).

Consumer Goods and Services

In general, purchases of many types of consumer goods have declined because they are frequently rented or borrowed from the Common Goods Stores at neighbourhood centres, or shared between family and friends. Gallup polls have recently shown that there is a strong sense of 'less is better' when it comes to individual ownership of non-essential products, though people are still interested in enjoying the use of them. Rental and shared use are seen as key to wise personal budgeting as well as to reducing energy consumption.

Appliances are now much more efficient than in 1990, and half of the stoves and clothes dryers and two-thirds of the hot water heaters are powered by natural gas. The results of these changes are shown in Figures 5.7 and 5.8. Appliance durability has also increased. Household furnishings such as furniture, floor coverings, and fixtures have decreased somewhat because people prefer dwellings that are easier and

Figure 5.4

Insulation levels of housing stock, 1990 and 2030

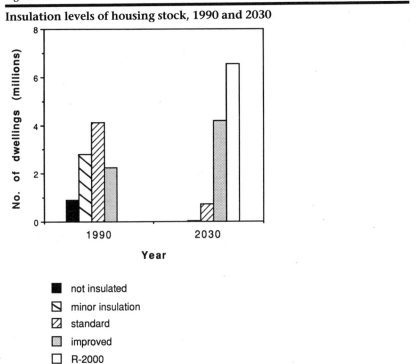

Figure 5.5

Home heating energy use, 1990-2030

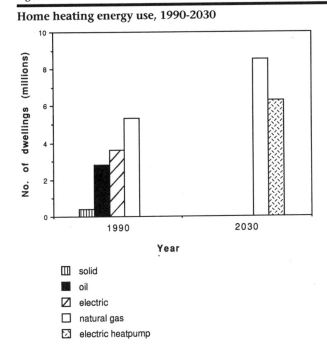

Figure 5.6

Household furnace types, 1990 and 2030

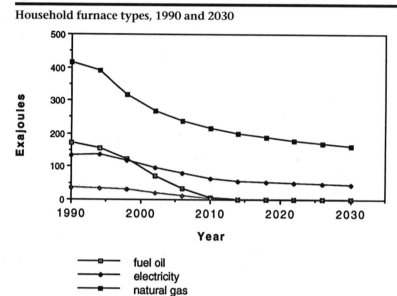

less expensive to maintain, and in general have become accustomed to smaller living spaces and the increased use of common areas. Purchases of recreational equipment such as electronic video and audio equipment have remained at late twentieth-century levels.

The decrease in dwelling size per person has also led to a decrease in the consumption of non-durable and semi-durable goods such as cleaning products, light bulbs, and much of the paraphernalia of maintaining a house. Goods are routinely designed to use fewer materials and less packaging and to be recyclable when they no longer function. The increased durability of many of these items has also contributed to reduced consumer demand.

The demand for personal services, including laundry, dry cleaning, and hair care, is slightly increased in 2030 compared with 1990. In general, the substances and methods used in all service sectors are environmentally non-destructive. Services that could not make this switch have not survived. But the service sector in general is thriving, because more people find time to develop skills that they enjoy using to supplement their incomes.

In 2030, the conserver society ethic has reduced the per capita consumption of clothing and footwear, which are now on average more durable. Used clothing stores and shoe repair shops are more common.

Figure 5.7

Number of kitchen and laundry appliances, 1990-2030

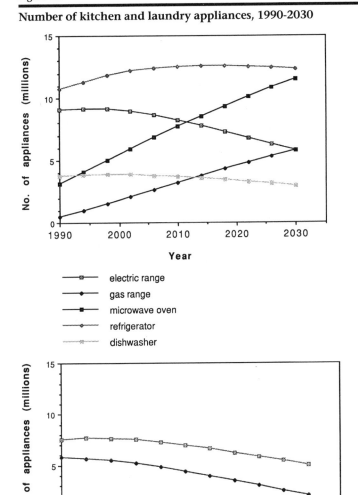

Personal adornment is still popular, but most people prefer to create their own styles imaginatively and inexpensively rather than invest heavily in packaged fashion. Demand for commercially produced toilette articles has also declined.

Figure 5.8

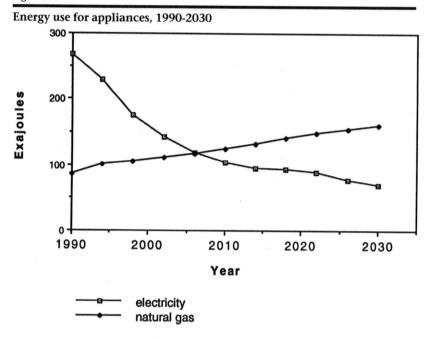

Energy use for appliances, 1990-2030

Legend:
- □ electricity
- ● natural gas

Health Care

Patients hospitalized for treatment of acute problems are now encouraged, where appropriate, to recover at home, as home care is generally believed to provide the best opportunity for the individual to recuperate successfully. The number of hospital employees per patient – other than doctors and nurses – has increased by 25 per cent between 1990 and 2030, with many of these employees engaged in a variety of home-care services as well as in some hospital attendance with long-term and hospice patients. In addition, many more visits are made to health professionals practising what was known in 1990 as 'alternative' medicinal techniques, including massage therapists, chiropractors, naturopaths, and acupuncturists, leading to a significant shift in the composition of the health care professions.

Average visits to doctors' offices (per age group) are 20 per cent fewer, but as a result of the aging population, and the fact that visits are now longer, there is a slight increase in the total number of doctors, especially for alternative forms of health care. The number of visits to hospitals by all age groups under sixty-five is down, with the length of stay up slightly (reflecting fewer but more serious visits). Even though average stays and the number of visits per person decreased between 1990 and 2030, the number of days in hospital per capita increased by 50 per cent. Because elderly people often require prolonged treatment for chronic conditions, large increases in special care institutions such as nursing homes have occurred. The number of days per capita at these institutions doubled between 1990 and 2030. The increases in numbers of patient days at all health care institutions, of course, meant large increases in numbers of employees. The health care sector grew to become a major employer in the Canadian economy, with over 1.6 million people employed in 2030, double the 1990 number.

Across all age groups, however, Canadians are healthier as a result of more leisurely lifestyles and healthier environments, including reduced urban air pollution. More nutritious diets, including reduced meat consumption, have also contributed to less demand for medical services. Because people are healthier both mentally and physically, the demand for prescription and non-prescription pharmaceutical products has decreased by 25 per cent across all age categories. Canada is also virtually tobacco free, and overall alcohol consumption is declining.

Education
Because education at all levels is now designed to develop Canadians who can engage in informed societal and environmental decision-making, most people have a relatively clear understanding of the types of constraints on human activities required by the imperative to preserve the biosphere. Many people are engaged in lifelong education, so they move in and out of the formal and the informal work forces to explore personal interests or simply to keep pace with the changing society.

Elementary and secondary schools are recognized as centres for developing sustainable society skills. Education for sustainability is now firmly entrenched in schools, colleges, and universities. Interactions that foster critical discussion are encouraged, and students are given the flexibility to experience both cooperative and self-directed learning relevant to their community and the world beyond. For at least forty hours per month beginning in Year Six, students assist in community sustainability projects in which they develop their organizational, social, and economic skills, and become familiar with community issues. This component of the educational experience allows students to exchange information and values with a wide variety of cultural and age groups in the community. College and university students and graduates constitute more than three-quarters of the population. These changes have resulted in a steady increase in postsecondary students, as shown in Figure 5.9. Teaching methods emphasize a team approach in which each teacher may have helpers who assist groups of students (e.g., many individuals over the age of sixty-five volunteer to help). There are generally high teacher-to-student ratios, so education is comparatively more labour intensive than it was forty years ago.

In 2030, parents are slightly less likely to start their four and five year olds in school, some preferring to keep them at home for schooling. In the six-to-sixteen-year-old age range, virtually all children are in a formal educational setting of some type. High school graduates increasingly travel and/or work before starting postsecondary education. Approximately 25 per cent of eighteen year olds are enrolled in colleges and universities (many are still travelling or trying out different kinds of work). Of those aged nineteen to twenty-four, 50 per cent are engaged in postsecondary education – 25 per cent in college and vocational institutes, 15 per cent full time in university, and 10 per cent part time in university. Part-time study is increasing, and twice as many people seek formal education at different periods throughout their lives, as was typical in the 1990s.

As a result of the low fertility rate between 1990 and 2030, the num-

Figure 5.9

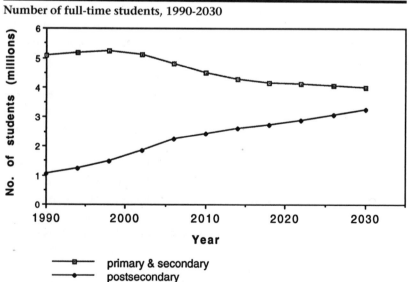

Number of full-time students, 1990-2030

Note: part-time students are counted as fractions of a full-time student

ber of primary and secondary school students decreased. The number of primary and secondary schoolteachers remained relatively constant between 1990 and 2030 due to an increase in teacher-student ratios. Despite decreases in traditional college- and university-aged people (ages nineteen to twenty-four), the number of students increased by a factor of three due to increased part-time postsecondary enrolment by adults of all ages. The number of instructors in universities and colleges also tripled over the scenario period.

Due to the increased teacher-student ratios, the increase in the number of teacher's aides, and the growth of postsecondary education between 1990 and 2030, the total number of persons employed in schools rose by 47 per cent. In 2030, over 1.1 million people work in the educational sector as either educators or support staff.

Media and Communications

Information plays a central role in maintaining societal sustainability in the Canada of 2030. Because feedback about the effects of all human activities on natural and social systems is essential, continuous information gathering, analysis, and dissemination are seen as particularly crucial. These programs have generous financial and political support, as well as the full cooperation of most private sector organizations.

The communications industry is quite different from that of forty years ago, because attitudes have changed about how broadcasting technology should be used. Broadcasting activities have increased fourfold from the 1990 level, and telephone usage has doubled in the same period. Much of the control of broadcasting is at the community level, with links to news and information from around the world. Daily personal electronic newspapers cater to a wide audience. The goals of publicly funded community broadcasting are to provide education, information about local activities, promotion of community development, and networking with other communities and regions.

The Canadian media provide comprehensive coverage of sustainability concerns on both local and global levels. In response to public demand this year, there have been a number of television programs and newspaper series detailing the progress finally being made in the transfer of 'clean' technologies to less industrialized countries. Advertising support from forward-looking industries that have 'greened' their products, processes, and labour relations is quite routine.

The increase in the number of combined broadcasting-telephone units (called superboxes) is counterbalanced by the decrease in cost per unit.[1] These superboxes combine the telecommunications technologies of the late twentieth century in one unit. They serve as combined fax machines, telephones, videophones, televisions, personal computers, and more. They have powerful interactive multimedia capabilities that allow users to transmit and receive news, literature, mail, and entertainment in video, text, and voice form.

In 2030, infrastructure (equipment and building) needs are 40 per cent less for both broadcasting and telephone networks than in 1990. Telephone services require 20 per cent less labour than in 1990, due to increases in efficiency. Because of the decentralization and proliferation of activities, as noted above, broadcasting activities require twice as much labour as in 1990, but this workload is offset by an increase in volunteers at community stations. The production of community video programs by amateurs has proven to be both feasible and relatively inexpensive.

The combined effects of the changes in the communications sector resulted in a 45 per cent increase in communications infrastructure and a 17 per cent increase in labour required between 1990 and 2030.

Food
Canadians in 2030 eat nutritional diets low in sugar and fat and high in fibre. Consumption of meat has decreased substantially (82 per cent between 1990 and 2030) because animal meat is high in fat, saturated fat, and cholesterol, and recommended dietary protein can be derived from sources such as beans, lentils, peas, and nuts or seeds (see Figure 5.10). In relation to 1990 levels, 13 per cent fewer dairy products are consumed, and 30 per cent fewer eggs are eaten. Overall, the past forty years have witnessed a trend away from processed foods. More people are eating whole-grain foods, and between 1990 and 2030, cereal consumption increased 26 per cent. A large increase in fruit and vegetable consumption (up 90 per cent from 1990 levels) has occurred because these foods are a good source of starch and fibre and can be purchased from local producers or processed at home. Based on 1990 food prices, the share of meat in the total food bill of the average Canadian shifted from 30 per cent in 1990 to 8 per cent in 2030, while the share of fruit, vegetables, and cereals grew from 24 per cent to 40 per cent of the total diet.

Recreation
Community theatre and cinematography groups are popular in 2030, and they promote the sharing of experiences between regions. The number of theatres and cinemas has increased by 15 per cent. The num-

Figure 5.10

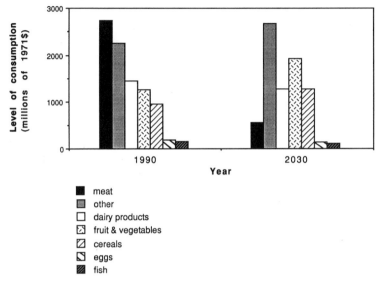

Food consumption, 1990 and 2030

ber of educational programs focusing on music, art, and other cultural services such as local museums has doubled in the last forty years. The shorter average work week allows more opportunity for people to participate in these leisure activities. The number of non-profit organizations staffed mostly by volunteers has increased substantially. The number of religious and quasi-religious organizations has remained constant, though there is an even greater variety of organizations with people exploring many paths to a meaningful spiritual life. The recreational use of parks and playgrounds has increased by 50 per cent because more people have access to open natural areas and other green spaces within a short distance from their homes. Community trails within and around urban areas are popular for hiking.

Recreation in 2030 is varied, and much of it is community based. This is partly a result of the high expense of travel, but also of increased involvement in one's own bioregion. There is strong interest in exploring the culture and landscape of one's own and nearby regions, with the trend toward simplicity and minimum expense. Small country inns, hostels, and restaurants have increased in number, as have picnic facilities. Luxury hotels have declined relative to more inexpensive accommodation, though some up-scale hotels are maintained. Demand for such facilities declined when tax-exempt business travel and entertainment stopped in 2002.

Transportation

Automobiles: Canada's transportation sector in 2030 suits its bioregions. The trend toward complete communities means that people don't need to travel as much as they did in 1990 for work, shopping, and recreation. Green space connects business, residential, and recreational areas. People walk or cycle in and around the urban core. Long-distance travel for recreation and business has decreased since 1990 because full-cost pricing of fuels makes such trips quite costly and more business is conducted electronically through video, telephone, and computer linkages.

In 2030, attitudes toward using automobiles are quite different than they were forty years ago. Seventy per cent of Canadian households have one small electric car, 10 per cent own two electric vehicles, and 20 per cent have no motorized vehicle. (In 1990, about the same number of households had one car, but the number of two- and three-car families was much higher.) There are now two types (though still many models) of private cars: about 80 per cent are small electric vehicles, ideally suited for urban driving, and the rest are larger – midsize by 1990 standards – fuel-burning cars for longer trips. The small electric cars are made primarily of a lightweight, durable plastic.

Half of these electrical cars are powered by batteries charged by the electrical grid; the other half are run by hydrogen fuel cells. Because the electric motor's energy conversion is far more efficient than that of the internal combustion engine (90 to 95 per cent compared with 12 to 30 per cent), these vehicles use much less energy than the gasoline automobiles that they replaced. Most intercity auto travel is by rented, standard-sized cars. These cars are similar in size to their compact predecessor of forty years ago, and are also made primarily of a durable, lightweight plastic (with a twenty-year half-life). There are two energy sources for the internal combustion engines of these standard-sized automobiles: clean-burning alcohol from forestry products, and hydrogen.

The total energy used for automobile transportation in 2030 is only 295 Petajoules (PJ). To travel the same distance, the automobiles used in 1990 would have required an estimated 573 PJ of gasoline. This fuel saving represents the improved energy conversion efficiency of the automobile engine, as well as continued improvements in automobile design features such as less aerodynamic drag and weight. The 1990 auto fleet actually consumed over 1,000 PJ of fuel. The additional fuel savings in 2030 have been realized by a combination of a 10 per cent decrease in the total auto fleet over 1990, a 16 per cent reduction in distance driven per vehicle, and an increase in the share of compact cars from 40 to 80 per cent.

Urban Public Transit: Urban public transit in 2030 remains a mix of motor buses, electric trolley cars, light rail (including monorail vehicles), and heavier rail (including subways). Alcohol-fuelled motor buses account for 60 per cent of urban transit vehicles, electric-powered trolley cars and light-rail transit for 15 per cent each, with the remainder heavy-rail transit such as subways. Motor buses powered by alcohol fuel are much cleaner burning than their diesel and gas predecessors. Concern for air quality in cities has resulted in more electric-powered street cars and light-rail transit systems. The electricity needed to power such systems is generated by hydro and other renewable energy sources.

The increasing attractiveness and utility of public-transit routes, combined with the changes in housing densities described above, have caused commuter trips to increase by a factor of 2.5 over the past forty years, as shown in Figure 5.11. Transit routes have more extensive coverage, and are used more efficiently. More stringent energy-efficiency standards have cut the energy used by public-transit vehicles by 50 per

Figure 5.11

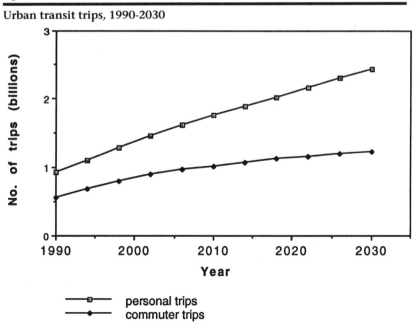

Urban transit trips, 1990-2030

cent from 1990, and, like autos, public-transit vehicles are more durable than they were forty years ago. In response to the decreased ownership of personal automobiles, taxi transportation has increased 25 per cent since 1990.

Other Public Transit: The use of the automobile for intercity travel has declined 15 to 30 per cent (depending on trip length), because more trips are now taken by rail and bus (see Figure 5.12). The Windsor-Montreal and Edmonton-Calgary transportation corridors, for example, are now served by high-speed electric trains. There has also been a modest shift from air travel to rail and bus, because these latter modes have become more pleasant and efficient and people have more time.

While long-distance bus-passenger kilometres increased 85 per cent between 1990 and 2030, energy use only increased 16 per cent due to improved energy efficiency and the optimization of bus scheduling. In 2030, half of the buses are powered by methanol and half use hydrogen. Passenger-train transportation has increased dramatically. These modal shifts have been accompanied by a 20 per cent decline in

Figure 5.12

Passenger transportation between cities, 1990-2030

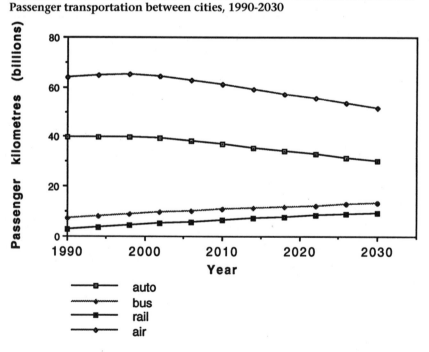

medium- and long-distance travel between cities, driven by higher costs due to the internalization of environmental costs in transportation prices together with telecommunications substitution.

Due to the uncertainty in the research of the alternative-fuel potential for airplanes, no changes in plane fuel were assumed for this study. The shift to ground-travel modes, however, allowed for a decrease of 19 per cent in air-passenger kilometres between 1990 and 2030.

Freight: There is a slightly decreased need for freight transport now because there is greater reliance on local resources, consistent with the overall trend toward greater regional self-sufficiency and local labour. As well, a larger proportion of the economic activity in Canada is now based in the service sector, with reduced need to transport either raw materials or manufactured goods.

Integration of ship, rail, and truck transport has resulted in substantial reduction of energy use compared with 1990. In the Great Lakes region, for example, containerized freight shipped in small, double-hulled ships is unloaded at many small ports onto adjacent rail lines for transportation outside the region, or directly into small delivery trucks for local distribution. This successful integration has led to a 50 per cent increase in ship dockages in the Great Lakes. Money spent on research and development to increase the fuel efficiency of ships has paid off substantially. Similar research on the use of trains for freight transport has led to an increase of 20 per cent, on average, in the distance travelled by freight trains. Because rail is used increasingly for long-distance freight transport, the amount of train stock and average train length have both increased.

In 2030, there are 40 per cent fewer trucks than there were in 1990, primarily because trucks are packed more efficiently and fewer trips are made with no load. The law restricting the transport of goods between provinces on return trips (e.g., between Ontario and Manitoba) was rescinded in 1998.

Government
In 2030, the structure of governments has changed significantly. In an effort to bring decision-making closer to the community affected, there has been a reorganization of responsibilities among the municipal, provincial, and federal governments. Detailed political management of cultural and environmental matters is more decentralized than it was in 1990, but takes place in the context of a stronger common legal

framework for environmental and cultural rights. The federal government now delegates management to lower jurisdictions in return for
legally binding agreements on basic environmental and cultural rights
and responsibilities at both the individual and community levels (see
Chapter 6).

By local community preference, the provinces are still responsible for
funding education and health services, while the communities administer these services. Other responsibilities formerly under provincial jurisdiction have become the responsibility of local governments. These
changes have meant that the number of administrative employees at
the municipal level has increased by 25 per cent since 1990, while administrative staff employed by the provincial governments have under-
gone a comparable decrease. Although employees of the federal
government serve a larger population in 2030, their number has remained constant due to an increase in efficiency.

Public transportation has seen a growth in the number of administrative employees, while the number of administrative employees at banks
and insurance companies has held constant at 1990 levels. There has
been no increase in administrative employees in national defence either, even though the role of the armed forces has changed significantly
to include environmental concerns. By 2030, former defence adminis-
trators are supervising a Conservation Corps that works to improve environmental conditions for all Canadians.

The result of the increase in the number of municipal government employees and the drop in provincial government workers was that the total number of government employees remained relatively constant,
while the population increased 10 per cent. More than 700,000 people
are employed in all levels of government in 2030. The public-transporta-
tion sector has seen a doubling of the number of employees to 30,000.

Office Sector
Office buildings, including those used by government, financial institutions, and insurance companies, are much more energy efficient, using 50 per cent less energy, on average, than those typical of the 1980s.
There has been a shift to increased electrification, with all remaining fuel
demands being met by natural gas. The average life span of office buildings has increased by 20 per cent, and there has been a general decrease
in the amount of goods and materials used in new office buildings.

While the total office-building stock remained constant between
1990 and 2030, the energy use dropped by 58 per cent. The fuel mix

stayed nearly constant, with two-thirds of fuel use as electricity and one-third as natural gas.

Retail Trade

The general decrease in the consumption of goods and services described in other sections of this chapter has decreased the need for storage space, which has declined 25 per cent per household. For the same reasons, wholesale capacity has dropped by 20 per cent per household. Because there are more small retail stores serving local communities, retail capacity per household has remained constant at 1990 levels. As a result, labour requirements per unit of capacity for storage and wholesale have remained constant since the 1980s, but have increased by 40 per cent for retail stores.

While the wholesale and storage components of the retail trade sector changed little over the scenario period, the retail component's increase in labour requirements resulted in significant changes. Because the retail sector is such a large employer, the increases resulted in the generation of approximately 1.7 million person years over the 1990 level.

Infrastructure

The modal shifts in transportation described above have led to significant changes in the transportation infrastructure. The total amount of transportation infrastructure in 2030 is 11 per cent lower than in 1990 for highways and roads, and 50 per cent higher for shipyards, 43 per cent higher for urban transit systems, and 17 per cent higher for railroads. Air-travel infrastructure has remained constant.

Municipal water- and sewage-treatment capacity expanded 30 per cent by the year 2000 to correct existing inadequacies in water-treatment facilities in Canada. Capacity then began to decline (to 1990 levels by 2030) because of more building-level water-treatment technology and a general decrease in water-pollution levels, both of which reduced the need for municipal water-treatment facilities. Water use per capita has decreased by 50 per cent since the 1980s because of more efficient use (e.g., waterless toilets, grey-water systems, improved industrial heat exchangers, etc.).

Manufacturing and Resources[2]

Due to reductions in the demand for many goods, total industrial activity decreased by 10 per cent between 1990 and 2030. A sectoral comparison of industrial activity levels reveals that activity in the mining, forestry, food and textiles, chemicals, and construction industries dropped slightly between 1990 and 2030 (see Figure 5.13). On the other hand, vehicle manufacturing increased over the scenario period as a result of efforts to replace the transportation systems of the 1990s with more environmentally benign technologies.

Figure 5.13

Sectoral industrial activity, 1990 and 2030

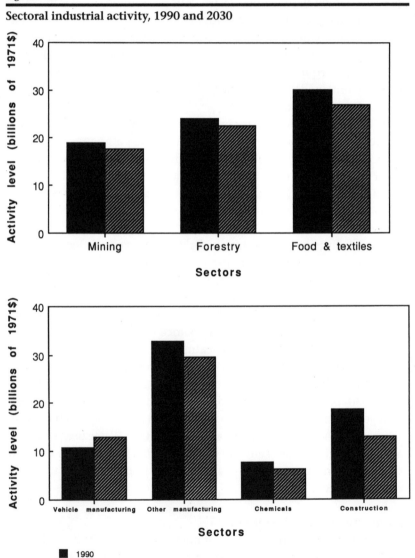

As a whole, the Canadian economy has continued its structural shift in the direction of more services related to manufacturing and resource production. While the resource sectors have also changed significantly in composition, such a shift has not occurred in the manufacturing sector to any great degree. However, manufacturing industries have become much more materials, energy, and labour efficient over the last forty years. On average, labour requirements per unit of output have continued their historical trend and decreased about 50 per cent by 2030, while capital stock per unit of output has, on average, decreased 25 per cent. These changes reflect increased efficiency in the use of inputs, including increased automation.

On the energy front, two major changes have occurred: significant improvements in energy efficiency, and a large growth in the cogeneration of electricity and process steam to replace the separate purchase of these two energy commodities.[3] On the efficiency side, energy used for motive drive has been reduced, on average, 20 per cent per unit of output from 1990 values. Energy for lighting has been reduced 67 per cent, improved combustion processes reduce steam and direct fuel use per unit of output by 20 per cent, and steam and direct heat used for pumping has been reduced 30 per cent due to heat recovery processes, which thereby increase electricity use by 10 per cent. With regard to cogeneration, 50 per cent of electricity demand is supplied by gas-fired cogeneration capacity, at a gas efficiency of 83 per cent. These savings are not additive but applied in the order described, resulting in a net saving for total energy use in manufacturing of about 28 per cent per unit of output.

The natural resources sectors have changed to ensure sustainable yields of precious commodities such as forestry, fisheries, and agricultural products. Despite some reductions in the harvest rates of these resources, Canada still meets much of its domestic needs, and continues to export many of the products of its renewable resource base. What follows is a description of the specific changes that occurred in the resource production sectors of the Canadian economy between 1990 and 2030.

Agriculture
The significant shift in Canadian agriculture between 1990 and 2030 has been a 15 per cent decrease in both crop-harvesting capacity and land farmed, coupled with a 15 per cent increase in livestock capacity and a 20 per cent decrease in feed-crop production. These changes reflect the return to grazing use (primarily in the Prairie region) of land

inherently unsuitable for crop production, and the general reduction of the size and capital intensiveness of farms, as organic farming methods eventually increased sustainable productivity. Fewer cattle are raised for domestic consumption in 2030, though exports of beef remain significant. Substantially more poultry are raised, primarily due to the increasing Canadian preference for poultry over beef.

The move toward sustainability in agriculture began with an awareness of the impacts of conventional agriculture. The transition to more sustainable, midsized systems of farming was by no means rapid. A gradual shift was recommended to allow farmers and farm economies, as well as agro-ecosystems, to adjust. The gradual introduction of (or return to) natural pest-control techniques and nutrient cycling has eliminated dependency on the synthetic chemicals of the twentieth century, and the use of appropriate technology has replaced energy-intensive high technology in the farm system. Wind, solar, and biomass energy are the main sources of power for the farm. The use of appropriate land-preparation techniques (tillage, irrigation, and nutrient-management practices) has reduced soil erosion significantly and protects soil quality. The volume of wastes generated is kept within the capacity of biotic systems to absorb them. These systems are flexible: farming methods are adapted to local ecological conditions, and on-farm inputs are substituted to the greatest extent possible for external inputs.

The severity and frequency of droughts in the Prairie region have increased moderately as a result of global climatic change, so that large areas of previously cultivated land were returned to rangeland and pasture, substituting for water-intensive grain production. In addition, large areas of wetlands have been restored under a comprehensive wetland conservation program. On the other hand, a significant reduction in the practice of summer fallowing has increased the amount of land available for crops and grazing. In the regions that can still support a variety of field crops, more drought-resistant plants (such as perennial grains) have been incorporated into less intensive, mixed cropping systems. The net effect of all these changes is a 15 per cent reduction in the agricultural land base, despite an increase in urban gardening.

In the last forty years, capital stock, including buildings, machinery, and other such materials, has decreased marginally by 10 per cent. Although the shift to smaller farms means that there are more of them and therefore more buildings per hectare, there is a net reduction in the amount of machinery required. Machines tend to be lighter and multipurpose, and used less frequently because of factors such as increased labour substitution and reduced tillage. Moreover, many farmers choose to share or rent farm machinery.

In 2030, Canada is somewhat less dependent on imports of food, a trend that began early in this century when California's Imperial Valley lost its water supply and no longer provided every conceivable fruit and vegetable all year long. The change in consumer expectations has led to the development of a highly nutritious and healthy diet utilizing primarily, but not solely, foods grown in Canada. Imports of feed crops also decreased sharply as grazing became more prevalent.

The production of livestock in 2030 exceeds the demand by a large margin. The demand for livestock products almost halved between 1990 and 2030, while the production increased by 15 per cent. Indeed, by 2030 Canada exports an average of 63 per cent of all the livestock produced, as shown in Figure 5.14. The production of crops also exceeds the domestic demand by large margins. The overall demand for agricultural crops remained relatively constant, while the production declined, as shown in Figure 5.15.

Overall, Canada continues to produce much more food than it consumes. Our scenario assumes continued availability of export markets

Figure 5.14

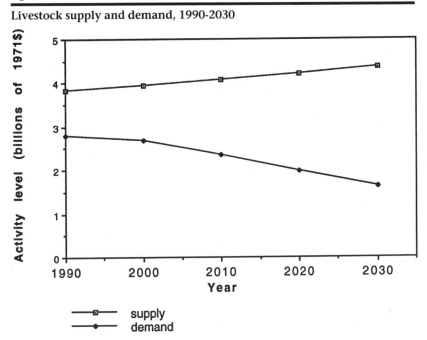

Livestock supply and demand, 1990-2030

Figure 5.15

Agricultural crops supply and demand, 1990-2030

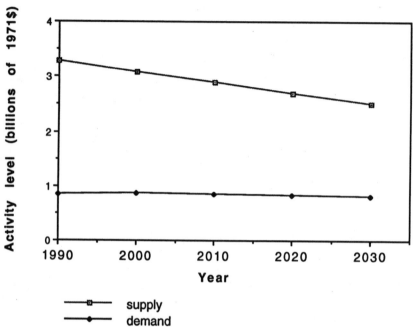

for surplus food production. This assumption holds for the other resource sectors as well.

The changes in the agricultural sector between 1990 and 2030 caused a 10 per cent decrease in the number of persons employed in this industry. This sector, however, remains a large source of employment, providing more than 520,000 jobs in 2030.

Fisheries

By 1995, Canada's inland stocks of native fish species, as well as its coastal commercial fish stocks, have been depleted to below regeneration capacity (fish reproduction and growth). An international agreement to prevent severe depletion of commercial fish stocks was reached in 1997, as were similar agreements for large inland lakes in the Prairies and Northwest Territories, in part due to the emergence of Native self-government. Commercial harvesting was reduced by 30 per cent over the following five-year period to allow fish stocks to regenerate. A shift to ecosystem management strategies, especially in the Great Lakes, and a steady decline in stocking inland waters with foreign fish species allowed native species to reestablish themselves in aquatic ecosystems, though sport fishing had to be heavily restricted. In 2030, the harvesting capacity still remains below the 1990 capacity by 10 per cent.

In 2030, there are efforts to reduce offshore fleets and to increase inshore fleets. These smaller scale, community-based fishing activities are characterized by environmental concern, sustainable 'common pool' management, and a 15 per cent reduction in the capital costs of catching fish. They are also 15 per cent more labour intensive, creating badly needed work for fishing communities.

Following the collapse of the east coast cod fishery in the early 1990s, the west coast commercial fishery was in serious trouble by the late 1990s. The high value salmon fishery was in decline due to activities in the interior of British Columbia that destroyed spawning beds. Commercial hatcheries were unable to replace the depleting stock. Pollution in the Fraser River and hydroelectric development on other rivers also adversely affected salmon runs. The halibut fishery was also in decline due to pressure from offshore fishing fleets that often entered the 200-mile limit. And the herring fishery was being overexploited as Canadian fishers responded to the incentive of high prices paid by Japan for herring roe. In the late 1990s, new regulations were introduced to reduce the size and catch capability of the fishing fleet, and to ensure that a larger share of the salmon fishery was allocated to Natives. Agreement was reached with Pacific Rim countries regarding the harvesting of halibut, and stiffer regulations were developed for the herring fishery.

In 2030, the ocean fisheries are gradually being restored to historical numbers, and provide 15 per cent more employment per unit of production than in 1990 due to more environmentally benign and labour-intensive fishing methods. Although the production level dropped between 1990 and 2000 to allow for the regeneration of fish populations, the supply of fish never went below domestic demand, as shown in Figure 5.16. With the introduction of more labour-intensive fishing technologies, the industry employs the same number of people in 2030 as it did in 1990, while production levels have dropped by 10 per cent.

Forestry
In 2030, having acted on the principle of sustainable forest yield, Canada is harvesting 50 per cent less forest than in 1990. Increased timber processing within Canada has reduced the number of raw logs ex-

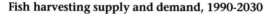

Figure 5.16

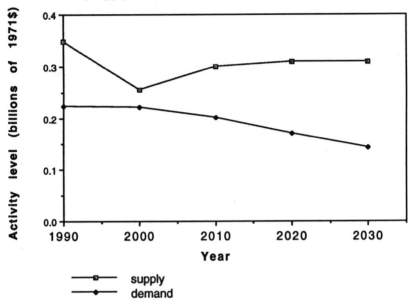

Fish harvesting supply and demand, 1990-2030

ported and has increased employment. Capital expenses to harvest a tree are 25 per cent fewer than in 1990, and new technology and stricter practices have made the forestry industry more environmentally benign.

Roundwood processing capacity in all categories has decreased. Pulp and paper production has decreased 50 per cent from 1990 due initially to recycling efforts in the 1990s and to the replacement of a large portion of print media by electronic media by 2015. Plywood/veneer capacity has decreased by 30 per cent. Population stabilization, accompanied by increased durability of the housing stock, has resulted in a reduction of 20 per cent in sawmill capacity.

Building waste is no longer automatically destined for the landfill, because buildings are dismantled in a way that allows many materials to be reused. Instead of large tracts of forest supplying a few large sawmills operating in northern Canada, there are smaller woodlots managed sustainably throughout the country that supply more and smaller local mills. The lumber for retrofitting existing housing stock and new construction is supplied mostly by these local operations.

The labour required per unit of production overall in the forestry industry has increased by 30 per cent since 1990 because of the importance placed on rehabilitation and the return to selective cutting rather than clear-cutting. As with all industry, there has been the adoption of the cradle-to-grave philosophy. Recycling paper products as many times as possible and planting trees to exceed those harvested are standard business practices. In the pulp and paper sector in particular, the emphasis on containing effluent has also meant that operations have become more capital intensive. This change illustrates how environmental costs previously externalized have become part of the cost of doing business.

Both the supply of, and demand for, forestry products have decreased between 1990 and 2030, and Canada continues to export large amounts of forestry products. Even with the 50 per cent decline in the pulp and paper industry over the scenario period, supply still exceeds domestic demand by a factor of 2.6 in 2030, as shown in Figure 5.17.

The forestry industry as a whole saw a slight increase in the demand for labour between 1990 and 2030 due to the increased labour require-

Figure 5.17

Forestry products supply and demand, 1990-2030

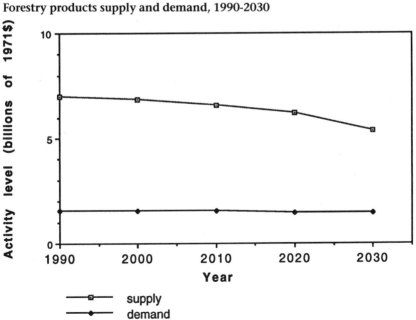

ments for forest rehabilitation and selective harvesting. In 2030, 244,000 Canadians are employed in this sector.

Mining

Mining processes have changed moderately since 1990, though employment has dropped 20 per cent in this sector because capacity has dropped and productivity has increased. However, the concept of cradle-to-grave responsibility – in the exploration and rehabilitation of mining sites, and in the collection of used equipment and scrap metals for reuse or recycling – has been incorporated into the planning and development of mining activities.

The infrastructure required for mining is different for gold, iron, base metals, asbestos, gypsum, salt, and sand and gravel than it was forty years ago. Labour productivity has steadily increased due to 'smart' technology and a managerial style that considers employee satisfaction and self-fulfilment.

Extracting, processing, and exporting uranium was phased out over the past forty years as the transition away from nuclear power was made. Uranium was used as a fuel source for nuclear reactors in 1990. Very few uses exist for uranium in 2030, though it is still valuable in some medical applications. Uranium for these uses is obtained from stockpiles dating back to 1990.

The demand for gold remains steady, because it is used in computer technology and a variety of other industries, as well as for decorative products such as jewellery. Iron production in 2030 has declined by 30 per cent due to decreased demand for steel in the fabrication of automobile bodies and other machines, and due to a general shift to a dematerializing of the Canadian economy and an increasing emphasis on services.

Mining activity for base metals (including nickel, copper, and aluminum) has held constant over the years to 2030. Ceramic increasingly replaces aluminum in engine blocks. The desired metal-refining capacity for aluminum has decreased by 50 per cent, because energy is no longer subsidized in Canada to promote the refining of bauxite shipped to Canada for processing. Asbestos has been replaced by more benign insulating materials because it posed a health risk and because fewer houses are built using gypsum; demand for it has declined by 25 per cent.

In 2030, the use of salt for de-icing roads has declined by 20 per cent because it was found to be corrosive of metals and harmful in large amounts to aquatic and plant life exposed to drainage from roads. The use of salt to flavour food has also declined because it was found to be harmful to health. Sand and gravel extraction and refining declined 25 per cent over the past forty years because fewer roads were built as urban sprawl was curtailed, and because fewer and smaller buildings were constructed, requiring less concrete.

Both the supply and the demand for mining products dropped approximately 20 per cent between 1990 and 2030, as shown in Figure 5.18. Canada continues to export large amounts of these products. It remains self-sufficient in the mining sector, and indeed continues to export large amounts of gypsum, salt, gold, iron, and other mining and base-metal products. Although employment in the mining sector dropped 20 per cent between 1990 and 2030, the industry still employs 152,000 people in 2030.

Energy Production
Total energy use has dropped dramatically. Much of the fuel used in 2030 is more environmentally benign than the fuels used in 1990, and is produced using renewable energy sources. Specifically, petroleum use

Figure 5.18

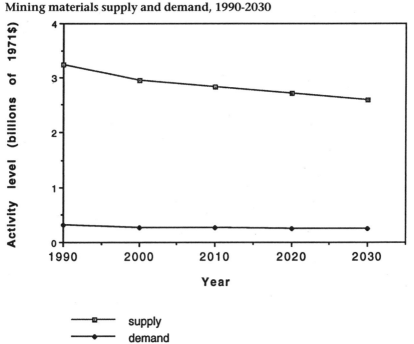

Mining materials supply and demand, 1990-2030

has dropped by 40 per cent, coal use has been halved, and electrical demand has dropped by 37 per cent, while the use of clean-burning natural gas has remained nearly constant (see Figure 5.19).

There has been a significant shift in energy supply patterns in Canada. Increased energy efficiency in all sectors has substantially reduced energy use per unit of activity. In addition, the internalization of environmental costs into energy prices has caused a shift away from non-renewable forms of energy, such as oil, gas, and coal, to several renewable sources of energy, which have witnessed continuing technological advances and, by the turn of the century, became economically competitive with conventional sources. The net effect of these two changes has been a decline in the use of conventional sources of energy

Figure 5.19

Energy use, 1990 and 2030

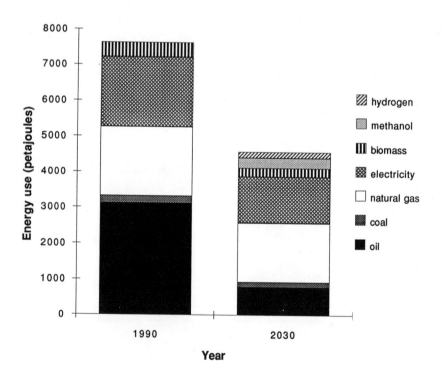

and the emergence of strong new industries connected with the delivery of energy efficiency and new renewable forms of energy.

Conventional Resources: Natural gas production in Canada has remained roughly constant from the 1990s to 2030, reflecting its relative abundance and the fact that it is the least environmentally destructive fossil fuel (see Figure 5.20). Conventional oil production, however, has dropped by 50 per cent from the early 1990s. This drop is a result of environmentally driven cost increases, increased scarcity in conventional areas of production, and significantly reduced exploration efforts, all of which have pushed prices up and made oil less competitive. The production of synthetic oil, which is abundant, has remained roughly constant. Improvements in materials efficiency similar to those in other

Figure 5.20

Primary energy production for domestic use, 1990 and 2030

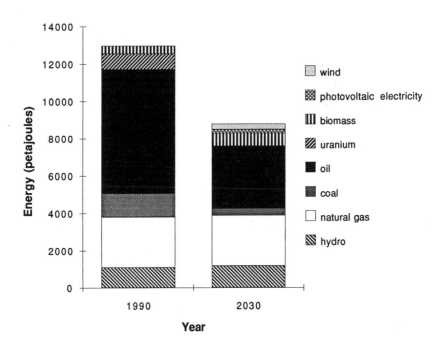

extractive industries have occurred, so goods and materials use per unit of oil or gas extracted has declined by 40 per cent.

Hydroelectricity production has increased only 10 per cent over 1990 levels. This minimal increase reflects environmental concerns over new hydraulic capacity, concerns that resulted in a virtual freeze on new developments after the year 2000, and in the cancellation of the James Bay II development.

Coal mining has declined 70 per cent from levels in the 1990s. This decrease is a direct result of the internalization of environmental costs in the price of fossil fuels. In particular, the carbon tax adopted in 1999 – after the UN Conference on Environment and Development in 1992 was followed by several years of record temperatures and a growing public outcry for response to global warming – made coal use prohibitively expensive. While clean coal technologies have made significant strides in regard to the reduction of other emissions, the continuing high cost and efficiency penalties associated with carbon-removal techniques have dictated a continuing decline in coal production.

The continuing high costs of nuclear power, and ongoing concern about safety and waste disposal, have effectively killed the nuclear industry in Canada. While progress was made on inherently safe reactor design in many countries during the 1990s, several significant accidents in existing reactors over that period contributed to increased public opposition to nuclear power in general. Growing stockpiles of nuclear waste, and continuing difficulties in getting public approval for waste disposal sites, exacerbated the difficulties. Moreover, the increasing efficiency of electricity and the cost effectiveness of new renewable forms of electricity production undercut the argument that nuclear power was needed to substitute for fossil-generated electricity. In Canada, no nuclear plant has been approved since the exorbitant Darlington plant was completed in the 1990s.

Renewable Energy Resources: On the renewable energy side, two new industries in particular have sprung up. The first is a biomass-based alcohol-fuel industry, which has two sources of feed stock: (1) the efficient use of forest and mill wastes, crop residues, and municipal wastes, and (2) to the extent required, high-yield plantations of hybrid poplars, grown entirely on abandoned farmland, of which there is 15 per cent more than in 1990, and on unimproved land. These plantations represent an 'agricultural' use of forest resources, and are cropped on an eight- to ten-year rotation.

Environmental groups in Canada have been very concerned about the long-term environmental implications of such an intensively managed land use. However, increases in vehicle-fuel efficiencies, coupled with a much more diverse set of sources for transportation energy, have kept required biomass production at a relatively modest level, thus minimizing the proportion needed from biomass plantations. Considerable care ensures as environmentally benign plantation management practices as possible. Nevertheless, there is a recognition that a biomass-based alcohol-fuel industry based on plantations must be transitional, buying time for the long-term transition to more environmentally benign and sustainable energy sources.

By 2030, about half of the over six million hectares of abandoned farmland in Canada in 1990 is being used to grow wood for the production of methanol fuel. These plantations are sufficient, on a sustained basis, to supply the yearly demand of 294 Petajoules of methanol required by Canadian vehicles.

The other new renewable energy industry is less problematic from an environmental point of view. Advances in the 1990s in thin-film technologies led to significant reductions in the cost of electricity from photovoltaic cells, which became fully competitive with utility electricity production from other sources just after 2000. This advance in turn opened the door to the economical production of hydrogen fuel from electrolysis on a large scale.

The availability of cost-effective electricity and hydrogen from solar energy occurred over the same period via significant improvements in battery and hydrogen-storage technologies. The combined result was the emergence of several new kinds of energy systems for vehicles, all based ultimately on photovoltaic electricity: hydrogen-fuelled vehicles, battery-powered vehicles, and electric vehicles using hydrogen-based fuel cells. This technology in turn led to a significant restructuring of (and growth in) the vehicle-manufacturing industry, and a much greater diversity in fuel types in the transportation sector.

On the energy-supply side, the main effect was the emergence of large solar-photovoltaic and hydrogen-production industries. In general, it is less expensive and more efficient to transport electricity than hydrogen, and this fact has required some increased interconnections between provinces to carry the resulting electricity. In some cases, however, the existence of pipeline rights-of-way and the technology made available by the decline in oil shipments, together with the specific end-use circumstances, have resulted in hydrogen-production facilities being constructed at the photovoltaic installations and in the movement of hydrogen by pipeline. By 2030, a network of 67,500 hectares of photovoltaic collectors is used to produce the 162 Petajoules of hydrogen required by the fleet of hydrogen-powered vehicles.

Electricity Production: The result of the changes in primary energy production noted above has been significant impacts on electricity production (see Figure 5.21). Nuclear energy has disappeared, while hydraulic power has increased only 10 per cent above 1990 levels. There has been significant growth in photovoltaic-electricity production, though little of this source of energy is fed into the grid because most is used directly for transportation purposes, as described above.

There has also been a significant growth in non-utility generation, both cogeneration systems used mainly for industrial applications and direct power production for sale to the grid. Much of the non-utility generation is natural gas based, using both combined-cycle gas-turbine

Figure 5.21

Sources used to generate electricity, 1990 and 2030

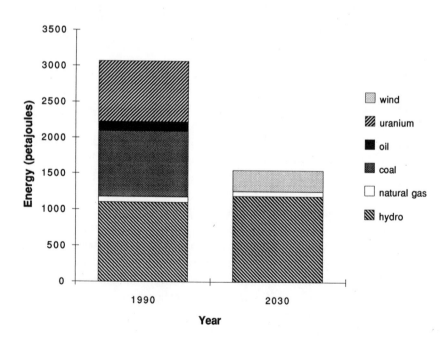

technologies applied widely in the 1990s and their successors. The rest is based on hydraulic power, often micro run-of-the-river hydro installations.

There is also a significant contribution to electricity production from intermittent renewables. Advances in wind-turbine design over the final decades of the last century continued into the new century, lowering costs and increasing the reliability of wind-power systems. The use of wind energy is constrained in Canada not by the physical availability of the resource, which is abundant, but by the extent to which intermittent wind energy can be accommodated on the grid without causing reliability problems.

The electrical production system of 2030 is dominated by hydraulic and wind sources. In 2030, 1,184 PJ of utility-generated electricity is produced hydraulically, and 296 PJ is supplied by wind-generating stations. These two sources supply a total of 1,480 PJ of electricity yearly, and are sufficient, except during peak demand times, to supply all of Canada's electrical needs, which are not met through cogeneration in industry. Continued electrical conservation efforts have helped to reduce the total electrical demand from 1,962 PJ in 1990 to 1,307 PJ in 2030.

A small fraction of electricity production, during peak demand times, comes from natural gas and biomass, using advanced combined-cycle turbine technologies in both cases, supplemented by fuel-cell technologies in recent decades. The electrical distribution system of 2030 takes advantage of a nationally interconnected grid to help mitigate regional supply and demand problems.

Part B: Institutional Dimensions

Political Decision-Making

Canada is now organized somewhat differently than in 1990 with respect to political jurisdiction. Increasing pressures for recognition of cultural and aboriginal differences, combined with a push to more localized and decentralized decision-making, have resulted in an increased devolution of powers to the various regions of Canada. Environmental concerns have led to an increased recognition of the need for coordinated goals and strategies and strong common standards on issues such as greenhouse gas emissions, the use of toxic chemicals, waste management, and the sustainability of resource development.

Detailed political management of cultural and environmental matters is even more decentralized than it was in 1990, but this takes place
in the context of a stronger common legal framework of environmental
and cultural rights. A partial version of this approach had been pio-
neered by Quebec separatists under the name of 'sovereignty-associa-
tion.' The generalization of this approach to environmental concerns,
combined with the growing recognition of the First Nations right to
self-determination and the growing global integration of economic activities, led to the idea of combining the devolution of management
with binding common rights and responsibilities. In Canada, this devolution is increasingly made beyond the provincial level to smaller regions and communities. An innovative form of the contract system has
been developed whereby larger political units expressly delegate management power to smaller jurisdictions in return for legally binding
agreements on basic environmental and cultural rights and responsibilities at both the individual and community levels.[4] This arrangement
allows for more direct political involvement by Canadian citizens on issues related to environmental management and cultural development,
at lower levels of spatial organization better attuned to immediate ecological experience and boundaries. Such a model is also more and more
common on the global level, as international agreements of this kind
are developed on issues ranging from global climatic change to common labour standards.

The establishment of binding agreements on basic environmental
and cultural rights and responsibilities required a process by which
such rights and responsibilities could be defined. A process of public
consultation at the national level (linked to international negotiations)
has therefore been established; through it, environmental strategies,
programs, goals, and targets can be proposed, debated, and decided on.
Such targets then become the basis of the agreements at the lower levels
of jurisdiction in exchange for the delegation of cultural and environmental management. The establishment of this process has been the
primary means by which environmental decision-making, broadly conceived, has come to be integrated into economic and social decision-
making at the national level. This development was intimately
connected with changes in the legal and economic context of decision-
making.

The combination of stricter environmental standards and targets
with a more comprehensive social welfare system has created an increased need for public revenues, and thus taxes have gone up signifi-

cantly. Moreover, the widespread implementation of effluent charges to internalize the environmental impacts of pollution has increased prices for many goods and services (though the revenues generated by the charges have offset tax requirements to some extent). The political will to enact such measures was generated by a public recognition that incurring such costs was necessary to avoid even greater environmental costs later, and to foster the development of industries based on clean and green technologies. Indeed, it became apparent late in the twentieth century that the development of such industries was a major route to maintaining industrial competitiveness in an increasingly integrated global economy.

More and more, national level politics is seen as limited to the development of common national standards, rights, and responsibilities in the economic, environmental, health, and cultural spheres; the collection and disbursement of government revenue; the management of the macroeconomy; and the management of international relations. The detailed management of resources or sectors is increasingly left to lower levels of jurisdiction, or to the marketplace. Although this management is constrained by the agreements and standards mentioned above, it represents a real increase in local authority, because the responsibility for determining how to meet the standards and targets has led to a burst of creative policy-making and entrepreneurialism at the local level.

The changes described here have been accommodated without major changes in the way that national and provincial politics are organized in Canadian society. Canada is still a parliamentary democracy in that parties are elected to govern at both the federal and provincial levels, though the authority of the provinces has declined relative to smaller jurisdictions. In contrast to the situation in Western Europe in the 1990s, for example, and in keeping with typical Canadian approaches, green politics has tended to be internalized within the existing parties. The political spectrum, however, has shifted toward a recognition of the need to respect certain environmental limits and to incorporate new approaches in economic and social decision-making. The main differences between the parties concern how to respect these limits and implement these approaches, ranging from the more market-based and biophysically oriented policies of the Progressive Conservationists through to the more community-based and socially oriented proposals of the Green Democratic Party (GDP).

More change is visible at regional and local levels, as increased decentralization together with the binding agreements have given more power to, and placed more constraints on, the political process. Innovative approaches to monitoring and enforcing national and provincial standards, often based more on local involvement and political, legal, and economic incentives than on top-down regulation, have provided the impetus whereby more environmentally benign and culturally sustainable behaviour is increasingly rewarded and internalized at the local level rather than imposed from above.

Most communities have bound themselves to monitor and meet certain environmental and cultural standards, and in return have received much more local authority over such matters. The result is a more highly charged, more powerful, but also more constrained local politics. As already demonstrated before 1990 at the provincial level, the devolution of powers to communities has led to more intercommunity bargaining and 'deal-making,' as each community attempts to balance its resources and capabilities in relation to other communities.

Thus, far from being a static sustainable political culture, Canada in 2030 is a country of creative tension, with an informed and politically active public determined to maintain a balance between private initiatives and environmental imperatives. The aim is to ensure a viable, balanced, productive, and self-reliant economy that provides the basic necessities of life for all citizens through a guaranteed annual income (GAI), and allows politicians at all levels to remain free from dependence on the support of the private sector corporations that once controlled the economy and thus the fate of politicians. A number of the political mechanisms in 2030, such as public financing of elections and mandated public representation on corporate boards, were developed to attain this goal.

Legal Decision-Making

By 2030, the legal system has become an important means of aiding the transition to a sustainable society in Canada. There are four main reasons for this. First, and most fundamentally, there now exists a powerful, constitutionally entrenched Environmental Bill of Rights that guarantees each citizen the right to a healthy environment and requires the government to ensure the continuation of such an environment through the establishment of targets, standards, and regulations. The Bill of Rights is a major legal enforcement tool for national, provincial, and local governments. It also provides the legal basis for citizen action to sue if standards are not met. Fewer restrictions on standing and class-action suits means that the threat of citizen action cannot be ignored by industry or government. A number of judgments at both lower courts and the Supreme Court of Canada have considered the definitions of the terms 'sustainability' and 'healthy environment.'

While the Environmental Bill of Rights provides the legal basis for the establishment of environmental targets and standards, it does not indicate what those standards and targets should be. This is accomplished by the second main legal development in the environmental field: a statutory process for establishing national environmental targets and standards. This target-setting process has six components: the determination of public preferences for broad socioeconomic development paths; the development of proposed standards and targets based on their ecological implications; the analysis of the technical and economic feasibility of meeting the different proposed standards and targets; the choice of the targets and standards to be imposed; the assessment of the social and environmental impacts associated with the proposed standards and targets; and the analysis of the policy and implementation requirements prescribed in the targets and standards.

The role and responsibilities of Environment Canada have also been changed significantly. Its management responsibilities over resources such as forests or inland waters have been moved to separate sectoral ministries, while the department has become more of a coordinating entity on environmental matters, much along the lines of the Department of Finance. Environment Canada has responsibility for managing the environmental target-setting process described above, which provides binding constraints on the activities of other departments and levels of government.

The third major legal innovation in Canada has been the development of a much stronger, statutory process of environmental assessment (EA). EA is the main vehicle for assessing whether proposed policies, plans, or projects conform to the requirements of the Environmental Bill of Rights and to the environmental standards and targets. In both cases, the EA process is based on a broad definition of environment that includes the social and cultural dimensions of human activities. Thus, the EA process has also become a major means of integrating sociopolitical issues into environmental decision-making.

This complex set of roles for EA has resulted in the establishment of a three-tier system of assessments. The first tier involves the assessment of the environmental and social impacts of the proposed standards and targets. The second tier is the evaluation of the environmental and social impacts of plans and policies intended to implement those standards and targets. The third tier is the evaluation of the environmental and social impacts in the planning of specific projects.

The fourth major legal innovation is the development of binding agreements on basic environmental and cultural rights and responsibilities between municipalities or regions on the one hand and provincial and/or federal governments on the other. These agreements are the basis for the delegation of management authority as discussed above. As with the EA and target-setting processes, these binding agreements address social and cultural, as well as narrow biophysical, issues. Municipalities or regional governments are to recognize certain individual rights and to abide by certain collective responsibilities defined by specific national and provincial environmental and cultural standards. Environmentally, these rights are grounded in the Environmental Bill of Rights; culturally, they are grounded in the Canadian Constitution and Bill of Rights, as amended in 2010. The responsibilities are partly derived from the national level targets and constraints emerging out of the environmental target-setting process.

The explicit recognition of both individual and collective rights in the agreements ensures that minority rights are protected against abuse as management of environmental and cultural issues is devolved to smaller and smaller jurisdictions. While such rights are protected at the national level in the Constitution and in the Environmental Bill of Rights, including them in the agreements means that the lower levels of government are also responsible for the active defence and promotion of such rights.

Environmental Conditions

The changes outlined above have led to some important achievements in the realm of environmental/ecological sustainability. The near universal application of 'clean technologies' and related measures to achieve maximum efficiency in the use of energy, water, and raw materials has resulted in major reductions of 'residuals' from industries and manufacturing firms. The toxic substance elimination programs cosponsored by governments and industrial associations have provided the financial and legal incentives needed for firms to redesign their product lines and/or production processes to meet the stringent zero-discharge standards promulgated by the first round of environmental standard setting in 2000. On balance, these measures have improved the profit margins for the private sector. The 'clean technology with environmental audit services' sector has grown enormously over the past thirty years, and now provides tens of thousands of shared jobs and nearly 15 per cent of the 'Net Environmentally Sustainable Production' index of the national economy.

The highly contentious 'healthy environment' legislation passed with the Environmental Bill of Rights enabled industrial leaders and consumers' groups to target and charge recalcitrant corporations. This legislation increases producer liabilities, reverses the 'burden of proof' requirements, and permits class-action challenges to any corporation that makes or uses any substance on the list of toxic and hazardous substances.

Consumer reaction to these changes was mixed. Those dependent on the old disposable consumer goods complained bitterly when they had to forego some products entirely, or for months or even years until sustainable substitutes were found. However, a growing recognition of the environmental and economic costs of waste disposal and of the savings to be attained by reducing inputs and increasing efficiency of use resulted in the gradual internalization of a 'reduce and reuse' ethic on the part of both industry and consumers. Recycling programs are now privatized, and producers are jointly responsible with consumers for the proper disposal of wastes. More effort is required from consumers to clean and sort items in order to have the right mix of materials for the wide range of 'resources from waste' product lines that these firms produce and sell. All goods are designed for more efficient reuse and recycling.

The main threats to our life support system in 2030 come from the 50- to 100-year-old disposal sites for nuclear wastes, from buried liquid-chemical wastes now partly solidified, and from groundwater contamination throughout southern Ontario and the Principality of Quebec. Some new technical processes have been developed to reduce the solidified chemical wastes, but the expense involved has slowed their application. There are also problems with the implementation of the Global Air Protection Treaty developed in 2001 by the former United Nations, and contaminated air masses still flow across Canada periodically.

The Endangered Landscapes Trust recently reported that about 75 per cent of the lands necessary for the enhancement and protection of biodiversity in Canada have been secured and are administered under a variety of arrangements, most involving some use of local volunteers. The Endangered Landscapes program, now in its fourth phase, is concentrating on the ecological rehabilitation of lands providing links between otherwise isolated and relatively undisturbed sites; many of the latter were set aside during the twentieth century as parks, wildlife refuges, or ecological reserves. The long-term goal is to have sufficient examples of the full range of biotic diversity (defined at the levels of ecosystems/landscape mosaics, natural communities, species, and populations of species of particular interest) thriving within each major ecoregion of the country. Research continues into the various ways of re-creating certain ecosystems/natural communities documented historically but extirpated before examples could be saved.

Debate over 'biodiversity' continues. Recent taxonomic studies of various taxa of invertebrates and important soil fungi suggest that they are poorly protected by the existing sites, and that more protected sites may be needed for these organisms alone; the recent proposal for spider mite conservation, however, received little public support. Canada also has problems in meeting its obligations under the World Conservation Union's 'World Conservation Strategy-5.' Under this agreement, countries must respond to global priorities for conserving biodiversity. Canada now has primary responsibility for maintaining viable populations of grizzly bears; conservation measures are required throughout the mountain ranges of the West and for about half of the Prairies, and there are many local complaints.

Climatic change continues to add uncertainty about the effectiveness of these various efforts. A number of forest ecosystems appear unable to adapt at the rate that these changes are occurring; extensive die-offs are found in ecosystems that could extend their range (such as the southern deciduous forests), as well as in those whose range will contract considerably (such as the northern boreal forests). The Endangered Landscapes Trust was able to acquire several small remnant sites within the refugium zone predicted for boreal forests after a Japanese-owned pulp company went bankrupt seven years ago. The Trust is also experimenting with transplants of threatened southern deciduous trees to areas several hundred kilometres north, where climatic change is expected to provide them with a congenial environment.

The Environmental Bill of Rights requires us to maintain or enhance the integrity of ecosystems. This remains a challenge. The Great Lakes continue to improve in terms of the slow flushing out or sediment trapping of the residual toxic contaminants, but the massive new shoreline condominium and related developments over the past thirty years have destroyed most of the scenic beauty and much of the biodiversity of the unique shoreline ecosystems. Many of the old open-pit mines, gravel pits, and industrial sites have been 'greened' to the point that they are now pleasant and useful. With the abandonment of farms in the increasingly arid regions of southern Saskatchewan in particular, some natural plant succession is occurring; there is still debate about what the land will eventually become and whether or not it should somehow be 'managed.'

The Supreme Court decision on 'ecosystem integrity' in 2012 had a curious effect. In the case comment by Chan, Morrison, and Singh (2025), it was noted that a leading academic biologist who was also a renowned 'reductionist' (even at that late date) presented evidence on behalf of Tarmac Development International to the effect that populations, not even species, are the highest biological entity; 'ecosystems' don't exist, and this is why there had been so much controversy about them. A conservation biologist called by the federal government to testify agreed that there is a human component to every ecosystem, and that the concept of 'ecosystem integrity' had a cultural as well as an ecological basis. The court ruled that because of this cultural component, 'integrity' was in the eye of the beholder; 'ecosystems' have no ontological status, and the issue is entirely one of human preferences concerning their surroundings.

People can certainly respond easily to preferences, and few prefer degraded environments. With the high level of environmental sensitivity that has existed, it was relatively easy to mobilize efforts to rehabilitate badly degraded areas regardless of their official label; people also responded to the notion of 'ecosystem health.' Landscape ecology and design now have a major role in environmental decisions. Environmental standards have been established to protect water, air, soil, and nutrient cycles, as well as to limit the use of fertilizers and pesticides that don't break down within twenty-four hours. The local monitoring of the 'State of Sustainability' makes use of indicators such as nutrient cycles (or evidence of nutrient 'leakages'), biodiversity, and the health of animals. At one time, it would have been called monitoring the integrity of ecosystems.

Notes

1 Since there were no superboxes in the list of goods tracked in the input/ouput table of the SERF model, an increase in the number of microwave ovens per household was used to simulate the gradual penetration of superboxes in Canadian households. Microwave ovens were chosen due to their similarity to superboxes in terms of goods and materials consumption.

2 In SERF, the manufacturing sector is represented by a large (150 x 450) input-

output table, from which the natural resource sectors have been separated, and connected to a trade calculator. The input-output table is driven by demands for manufactured goods emerging from all other sectors of the economy, and determines the levels of manufacturing activity and resource production required to meet those demands. In turn, required levels of resource production are compared with planned levels, as determined by the input assumptions described in other sections of this chapter.

3 Cogeneration involves the simultaneous production of electricity and process heat using a single energy source (e.g., natural gas or biomass). It represents a significant saving of energy required to produce the same amount of electricity and process heat from different sources in separate processes.

4 In other words, Canada has moved, over the period 1990-2030, toward the same kind of loose confederation of partly sovereign political entities to which Europe was heading in the 1990s. The difference is that Europe was coming from the opposite direction, that is, from a set of already sovereign states.

6
Achieving a Sustainable Society

Sally Lerner, D. Scott Slocombe, George Francis, and John B. Robinson

Introduction

This chapter provides an overview of the policy implications of the results of the Sustainable Society Project (SSP). This overview includes the lessons of the computer model's testing of the qualitative and quantitative scenarios, supplemented by more conceptual and implementation-oriented background; some general conclusions derived from the project for fostering and implementing a sustainable society; and specific conclusions on what needs to be done next in support of such goals. The focus in this chapter is on policy implications today: what we would need to begin to do now to start the process of realizing a sustainable society in the future.

Policy is a general plan or course of action, or a principle, adopted by government, corporate, or other organizations. It is intended to ensure certain broad, desirable outcomes without precise or heavy-handed specification of what those outcomes are. It is closely linked to administration, the management of human affairs on a more detailed basis, but this chapter focuses on policy with only occasional forays into the specifics of administration.

Developing policy is a complex, subjective activity. There may be many routes to a particular sustainability goal. A first assumption is that an industrialized market economy and a parliamentary democracy will continue in Canada, albeit with major changes in the role of the market and in the forms of political and social organization. Sustainability goals may be reached by very different kinds of policies, even to the extreme of conceiving sustainable dictatorial societies as well as sustainable democratic ones. This diversity was recognized early in the SSP, and the project's fundamental values were partly selected to ensure that non-democratic societies were not considered appropriate. It is worth repeating those fundamental values here, because they deeply affected our scenarios and the policies that we suggest.

- The continued existence of the natural world is inherently good. The natural world and its component life forms, and its ability to regenerate itself through its own natural evolution, have intrinsic value.
- Cultural sustainability depends on the ability of a society to claim the loyalty of its adherents through the propagation of a set of values that are acceptable to the populace and through the provision of sociopolitical institutions that make the realization of those values possible.

Chapter 5 presents a detailed review of the implications of the SSP's qualitative and quantitative assessments and models for the whole range of socioeconomic sectors of society. In so doing, it demonstrates that at least one scenario exists that is physically and technologically consistent with the values and design criteria outlined in Chapters 3 and 4. As discussed in Chapter 3, however, the modelling approach that we used focuses on material flows and technological factors, and does not address many of the economic factors relevant to the values and principles of a sustainable society. These concerns can only be addressed by the development of policies that are sensitive to equity and distributive justice.

This chapter examines the ways in which government and corporate policy at all levels might contribute to bringing about a sustainable society. The focus on policy is critical because we cannot hope to specify exactly how a sustainable society will work, or what form it will take. We must seek to create the processes of governance and communication that will make such a transition inevitable, or at least probable, and self-sustaining.

Policy Principles for Sustainability

There are a number of specific, implementation-related principles that emerge and reemerge in the SSP. These principles must be manifested in policy for all sectors and levels of Canadian society. They are introduced in general terms in this section, and then illustrated in detail in the next section, with a focus on different levels and sectors. These principles are rights and responsibilities, participation, target setting, assessment, recognizing economies as subsets of ecosystems, and monitoring. Some of them have been discussed briefly in the last section of Chapter 5, while others are new. Each is addressed in turn below.

Rights and Responsibilities

Here the emphasis is on enshrining certain rights of the environment and of people with respect to the environment, while also defining and fostering recognition of the responsibilities of individuals and organizations to work toward sustainability.

As noted and discussed in Chapter 5, what is needed is a powerful, constitutionally entrenched Environmental Bill of Rights that guarantees each citizen the right to a healthy environment and requires the government to ensure the continuation of such an environment. Such a bill would serve as a major legal enforcement tool for national, provincial, and local governments. It would also provide the legal basis for citizen action by providing the right to sue both polluters and governments responsible for enforcement if standards were not met. Fewer restrictions on standing and class-action suits and the prevention of countersuits aimed at discouraging citizen action would also mean that the threat of citizen action could not be ignored by industry or government.

Participation

Participation is at once a right, a responsibility, and a process. The first must be addressed in an Environmental Bill of Rights, and should recognize two related principles of participation. The first is that individuals affected by economic and other decisions have a right to participate in and influence those decisions. The second is that individuals in a region have the right to participate in and influence decisions affecting their region's resources and environment.

The responsibility of participation is to inform oneself about the issues and alternative viewpoints, always trying to move beyond one-sided, short-term perspectives. The relevant information should be widely available, and education should be oriented toward ensuring that this information is understood.

Developing effective, efficient processes for participation will be a challenge. Some formal processes for participation might involve computer-mediated systems such as electronic bulletin boards or voting, hearings and information sessions, and local and regional decision-making councils. Participation, as defined and recommended here, is a strong force for the devolution of management from upper-tier to local authorities.

Target Setting

By target setting, we mean a statutory process for establishing environmental standards and targets. Target setting would build on improved processes of ascertaining rights and responsibilities through participation, and contribute to new statements of rights and responsibilities in, for example, an Environmental Bill of Rights. The process would also require an unprecedented degree of cooperation between levels of government, and between departments within levels. Such a process would have six main steps:

(1) the development of proposed standards and targets based on their ability to meet desired ecological objectives
(2) the determination of public preferences for broad socioeconomic development paths within the context of such targets and standards
(3) the analysis of the technical and economic feasibility of meeting the different proposed standards and targets
(4) the choice of the targets and standards to be imposed
(5) the assessment of the social and environmental impacts associated with the proposed standards and targets
(6) the analysis of the policy and implementation requirements implied in the targets and standards.

The process outlined here implies that a concrete link be made between sociopolitical and economic preferences (in step 2) and environmental constraints (in step 1). It would also require the development of innovative forms of linking science with public involvement and policy formulation. For example, steps 1, 4, 5, and 6 all require forms of public participation that go beyond the typical experience of the twentieth century.

Assessment

The need is for a statute-based system of environmental assessment (EA) that would be stronger than Canada's current fragmented and often discretionary procedures. Such a system would ensure the conformity of policies, plans, or projects to an Environmental Bill of Rights and to broader environmental standards and targets resulting from the target-setting process. The EA process must be based on a broad definition of environment that includes the social and cultural dimensions of human activities, so that it can be a major means of integrating sociopolitical, economic, and environmental factors in decision-making.

EA can be idealized as a three-tier system. The first tier involves the assessment of environmental and social impacts of proposed standards and targets emerging from target-setting processes. The second tier is the evaluation of the environmental and social impacts of plans and policies intended to implement those standards and targets. And the third tier is the evaluation of environmental and sociocultural impacts during the planning of specific projects.

The fundamental criterion of evaluation at all three levels of EA must be that any standard, plan, or project should result in no net decrease in environmental quality or health. In addition, proponents should be required to attempt to improve environmental health. These criteria, which should be rooted in the definition of a healthy environment in an Environmental Bill of Rights, would normally be assessed in terms of the four objectives/criteria outlined there as well. In addition, the second and third tiers of EA should build on the first. The process would likely be iterative and participatory, moving from national targets and standards to the local, then to local assessment, and finally back up the tiers of EA if the lower level findings require the reappraisal of the higher tiers.

Recognizing Economies as Subsets of Ecosystems
The conceptual key to integrating environment and economy is recognizing that economic systems are subsets of ecological systems. This means an end to decision-making that separates the economic and environmental factors relevant to the decision; it means recognizing environmental constraints as primary. The recognition of this primacy must be integrated into decision-making at all levels from the individual to the federal. Such integration is a critical aim and corollary of the first four principles: rights and responsibilities, participation, target setting, and assessment.

There are many ways to foster the integration of environmental and economic decision-making. State-of-the-environment reporting provides ongoing assessment of environmental conditions, threats to sustainability, and management actions. Natural resource accounting seeks to track the economic costs of resource consumption (depletion) in calculating GNP and the like. Real-cost pricing means including the costs of replacing consumed resources and cleaning up pollution arising from production in the calculation of a product's price. Cradle-to-grave management and accountability imply producer responsibility for the impacts of producing a product, from accessing the raw material to disposing the waste associated with the product. Incentives of various types are seen by some as 'carrots' that can be effectively combined with regulations to foster more realistic decision-making (Hawken 1993).

Monitoring

Monitoring implies detailed, specific tracking of economic, environmental, and social data and trends at all levels from the local to the international. It means knowing the state of the environment, the state of the economy, and the state of society. As far as possible, it means adding current data to an existing database of historical information. The effects of actions and policies are evaluated both to ensure sustainability and to learn from experience.

Innovative approaches to monitoring and enforcing national and provincial targets are needed. These should be based more on local participation and political, legal, and economic incentives than on top-down regulations. These approaches would provide the impetus for behaviour that is more environmentally benign and culturally sustainable and is rewarded and internalized at the local level rather than imposed from above. Monitoring is tightly linked to target setting, defining rights and responsibilities, and fostering participation, because they will determine what must be monitored.

Much monitoring could, and should, take place at local and regional levels. Communities could agree to establish and meet certain environmental and cultural standards, and in return receive much more local management authority. Environment Canada's recent effort to develop regional Ecological Science Centres for state-of-the-environment reporting is to be commended in this regard.

Well-funded Public Interest Research Panels, operating independently of the government and the private sector, could eventually produce annual 'State of Societal Sustainability' reports. This program would build on, and ultimately incorporate, existing state-of-the-environment reporting. These panels could be composed of representatives of citizen interest groups and professionals with required technical skills. They would be salaried, have limited terms, and report to the public as well as to an Office of the Defender of the Environment, which would scrutinize and evaluate the panels' work. This scheme could be linked to a citizens' Corporate Watchdog Movement, which would gather information on corporate behaviour in Canada and abroad, and publicize its findings widely. This information could be used effectively by consumer boycott groups, especially when corporate interconnections are recognized and understood.

2030: Policies and Practices for Getting from Here to There

The SSP goals, methods, and outputs are clearly relevant for communities seeking to chart a future for themselves that is both economically viable and environmentally sustainable. The framework of the SSP, especially the adoption of backcasting and scenario analysis as basic approaches, can serve as a model for communities that want to take their futures more firmly into their own hands.

Backcasting at the community level requires the development of a proactive, participatory process that involves a broad cross-section of residents in creating vision of the kind of community they want for themselves and future generations. From this initial exercise, they can derive broad principles for future development, a set of ideal characteristics for various community activities and components, and detailed design guidelines for their community as they want it to be at some future date.

Eventually, a community can move forward by creating future histories (i.e., revisioning on the basis of the initial steps) that clarify the choices and strategies required to get from 'Community Now' to 'Community Future,' specifying concrete objectives based on this analysis and on the design guidelines, and creating task forces to pursue these objectives with firm dates for reporting. As in the SSP itself, these activities will be iterative to provide for maximum input into the final community plan.

The particular value of this type of process is that it requires consideration of a variety of human needs and quality-of-life issues. Broad participation and iterative discussion ensure that conflicts in value are aired, that the resultant plans for the community's future reflect some measure of consensus rather than simply conventional wisdom ('bring in a megaproject') or the priorities of an elite, and, therefore, that the community as a whole 'takes ownership' of policy changes and works effectively to realize goals.

Population, Education, and Economy

A sustainable Canadian society, as described in detail earlier in this book, would have a diverse population with varied family structures and changing work patterns, trends that are evident now. Governments need to begin to consider what timely policy initiatives will be required to ensure that responses to these changed circumstances produce socially and environmentally sustainable outcomes.

For example, some form of guaranteed annual income will be a virtual necessity because secure, full-time, adequately waged employment will no longer be available to all who want and need it, due to technological change and a globalized economy. The guaranteed income, combined with the sharing of paid work, would support self-employment and child care as well as foster lifelong learning – from travel to skills development and retraining when desired or needed. Concomitantly, policymakers should look toward assuring accessible, affordable housing and transportation through densification and increases in shared space, as well as investment in public transit. High-quality health care and education should be focal points of public policy, and assurances should be provided that they will continue to be universally available to Canadians. Imaginative ways of producing revenue must become the goals of financial policy developers. They need to consider new taxes on 'bads' such as pollution and non-productive activities such as currency speculation, and on new technologies, perhaps in the form of a small levy on all electronic transactions.

From the point of view of the individual consumer, lifestyle changes would have two main origins. First, policy should dictate that the cost of consumption and investment behaviours that result in negative environmental impacts would increase significantly, as socially determined environmental costs are internalized in the prices of goods, resources, and services. Second, certain consumption and investment behaviours would be mandated or proscribed, though this proscription would often be accompanied by subsidies directed toward the replacement of existing environmentally damaging equipment and capital. These developments would take place within the context of a significant shift in consumer preferences toward more environmentally benign products and practices, a shift supported by educational and promotional activities that supplement market-based approaches in the manner suggested above.

Because people on lower incomes tend to spend a higher proportion of their income directly on environmentally derived services such as energy or food, the price increases associated with the internalization of environmental costs would have significant equity effects. These effects would be an important part of the rationale for the increased social welfare measures (e.g., GAI) discussed above. These measures would be supplemented by equity-enhancing schemes such as lifetime rates and a guaranteed minimum availability of essential environmental services. The cost of these latter measures would be factored back into the adjustments of market prices.

The overall result of these developments would be a significant shift away from the conspicuous consumption typical of the late twentieth century toward a consumption ethic characterized by thrift, value for money, durability, environmental healthiness, and waste reduction.

When this change is combined with the changes in political culture described here, it should be possible that, for most Canadians, the basis for satisfaction or self-fulfilment could shift to some degree from the realm of material consumption to that of political involvement. As people become more empowered in choosing their futures and in collectively deciding on the environmental and cultural context of their lives, they should find it less necessary to express themselves through the consumption of material goods.

A key component of this transformation would be education. Canadian education at all levels must be designed to develop citizens who can engage in informed societal and environmental decision-making. If this were so, most people would have a relatively clear understanding of the types of constraints on human activities required by the imperative to preserve the biosphere. They would also recognize the continual temptation for some individuals and organizations in both the private and the public sectors to throw off the 'heavy hand' of government sustainability standards and to go beyond the mandated constraints.

Resource and Economic Sectors

Policy about renewable and non-renewable resources would be significantly affected by efforts to foster a transition to a sustainable society. Policy-making must address goals such as the reduced consumption of meat and processed foods, and the increased consumption of grains and cereals; the move to more sustainable, medium-scale operations, natural pest control, and nutrient recycling; the ability to cope with climatic warming; the move toward genuinely sustainable, locally based forestry and fishery operations; and reductions in the consumption of non-renewable resources (detailed in Chapter 4).

Generally, the changes between now and 2030 will be a reflection of shifts to durability, self-reliance, conviviality, and non-material pleasures. The most important changes will be in the planning and management of the natural resources and economic sectors. The principles and policies outlined earlier in this chapter will provide a new set of priorities, criteria, and processes for the detailed science and practice of resource and economic management.

Thus, changes will be required to resource management policy based in rights and responsibilities, target setting, and public participation in order to complement changes in consumption and production patterns (Young 1992). For example, if management authority over forestry were devolved to a local community, this would imply the existence of an agreement on the part of the community to apply federal and provincial standards on sustainable yield, allowable types of forest management, permissible emissions, etc. No devolution of management would take place in the absence of such an agreement. Less capital-intensive, more labour-intensive and environmentally friendly harvesting technologies would be required, along with greater domestic or even local processing of timber. Pulp and paper production should be reduced through increased recycling and the use of electronic media.

Information and communication must play a strong role in maintaining environmental and societal sustainability in the Canada of 2030, especially so in society's use of natural resources for economic ends. Because feedback about the effects of all human activities on natural and social systems is essential, policymakers must mandate continuous information gathering, analysis, and dissemination. Monitoring programs, as described above, must have generous financial and political support, as well as the full cooperation of most private sector organizations. They need to be staffed by both paid professionals and, on a wide scale, trained volunteers. Healthy public respect for the limitations of even the best monitoring systems and wide acceptance – among decisionmakers as well – that caution and moderation in all human activities should be the norm are critical. A sustainable society would have resource and economic activities, just as a non-sustainable society would; the difference would lie in the processes, levels, and results of development and planning.

Conclusions

Several clear lessons emerge from the detailed exploration of scenarios and options in the Sustainable Society Project. These lessons reflect the principles and policy implications identified above.

(1) Decisions must be based on, and judged in terms of, sustainability, not special interests. This means using good ecological and economic information; developing alternatives and assessing options in participatory, open forums; and making decisions for the long term.

(2) Basic standards must be maintained while decentralizing the provision of services, decision-making, and information collection.

(3) People must be educated and employed in ways that provide for their active participation in the development of plans and the assessment of options for sustainability. This means assuring security for all through shared work and a guaranteed annual income.

(4) The individual and the community must be empowered to achieve collective as well as individual benefits. This could lead to more highly charged and powerful local politics that are simultaneously more constrained by rights, responsibilities, and targets defined at the local, provincial, and national levels.

(5) Government intervention must foster change, not maintain entrenched interests and obstacles to sustainability. Policy must be used creatively and proactively.

As we observed at the start of this chapter, large changes will only come about through the accumulation of many small ones. The key to a successful transition to a more sustainable society is therefore the development of a new policy framework that fosters and supports appropriate lower level initiatives.

References

Hawken, P. 1993. *The Ecology of Commerce*. New York: HarperCollins

Young, M.D. 1992. *Sustainable Investment and Resource Use: Equity, Environmental Integrity and Economic Efficiency*. Carnforth, UK: Parthenon Publishing

7
A Retrospective
George Francis

The Sustainable Society Project (SSP) was a stimulating intellectual exploration for those who became involved with it. The self-imposed challenge to think through sustainability was a powerful heuristic. It required examining and reexamining some basic assumptions and values about people and society. Heroic as some of the assumptions were, the process of addressing them dissolved many of the barriers that impede thinking outside of disciplinary 'silos' and the private preserves of professions. It required challenging some socioeconomic and other orthodoxies of the day, including some from 'environmentalists.' Discussions revealed both hopes and fears about the future. And to reiterate, the results from the exercise are not forecasts, only insights into what could be. A vignette of our scenario would have a Canada in 2030 that includes the following points.

- People accept a healthy lifestyle, as indicated by their personal habits (nutrition, physical exercise) and social and community relations.
- A guaranteed annual income combined with work sharing releases people for community involvement on a volunteer basis.
- Consumerism is considerably reduced; durable goods are more often shared or rented when needed, and are repaired or recycled when necessary.
- Education and training are both lifelong and varied in the ways that opportunities for them are provided.
- Urban communities have higher densities of somewhat smaller dwellings, more shared spaces within or between dwellings, and energy efficient designs.
- Public transit plays a considerably larger role in the transportation system, while cars are smaller, lighter, and more fuel efficient.

- Communication services and technologies have proliferated, especially at the household level, and they offset travel to some extent.
- Government services are closer to the community, but are subject to strong national and/or regional and provincial standards.
- Health policies emphasize preventive measures, alternative medicine is more common, and home care is provided to the extent possible for those in need.
- Renewable energy provides almost all of the energy needs, and energy efficiency pervades the economy.
- There is less farmland but more rangeland, more lower input organic farming, and more vegetarianism generally.
- Industries have taken on cradle-to-grave responsibility for their manufactured products, which in turn are more durable and reusable.
- Resources are used sustainably on the most cost-effective basis, and adapted to the ecological requirements of the different bioregions.

Several aspects of the SSP process stand out and are worth commenting on, as an early retrospective. The SERF model was both a strength and a constraint for the SSP. Its strength was its grounding in the material bases of the Canadian economy, linked to demographic dynamics and population driven. It forced attention to details easily glossed over by those who enjoy big picture views of the world and debate about more abstract issues. It created 'tensions' as reasonable sounding ideas about change clashed with current 'realities' built into its models, and it required deliberate choices among options. It was something that had to be navigated through, not something that could just be left to 'run' on its own. That one reasonably coherent scenario could pass this test was one of our prouder accomplishments. The 'results' were preferences that survived this basic test of their feasibility.

But like all models, SERF has its limitations. As an aggregate of the people, material, and energy flows through the Canadian economy, its use implies that Canada, conceived as a nation-state, is both an appropriate and a viable unit for analysis. While many of us may like to believe this, the forces driving change operate at all levels from global to local, and interventions on behalf of sustainability would have to match this range. SERF also replicates a sectoral view of the economy, necessitated in part by the available data. Thus, some intersectoral or other more 'holistic' linkages may be overlooked when trying to develop a scenario. SERF also could not capture essential flows that are not part of the material base of the economy, whether from environmental or human sources. These other aspects of sustainability, associated with ecosystem integrity, economic vitality, community well-being, and social equity, had to be brought in from outside SERF, and appropriate links to SERF variables were not always found.

Whose values are in the SSP? We could typecast ourselves as a small group of educated middle-class professionals, having some bias toward concerns about environmental and ecological sustainability. Our values are 'culture bound,' like others, but we expect many of them to be widely shared within the dominant cultures of Canada. By making them explicit, disagreements based on major philosophical differences can be identified and perhaps addressed at this level. Otherwise, they could surface as quibbles over details that never quite reveal the sources of the contention.

Is a sustainable society in Canada with just thirty million people completely out of the question? We chose thirty million because if our vision of a preferred future was not feasible for people already here or about to arrive, then we would have had to think in the context of very tight margins, near thresholds, and less flexibility for adjustments. The metaphor of the 'ecological footprint' points to the non-sustainability of current populations and lifestyles if Canada had to achieve sustainability endogenously in some completely self-reliant economy. Whether our scenario sufficiently modified lifestyles to offset this problem is a moot point.

The assumption about achieving endogenous sustainability was linked to the assumption that other countries, notably the United States, would be moving in the same direction. We thought that these assumptions were necessary to help bind the scope of the exploration, while being able to link it with SERF. Variables in SERF also took into account trade-related factors. But this is a difficult assumption to maintain. Canada is open and vulnerable to global changes, some of which can provide opportunities to enhance economic vitality, perhaps through the export of sustainability enhancing technologies, while others can accelerate environmental and resource exploitation within Canada. While our scenario does allow for the cessation of food imports from California, it does not try to address the net balance in 'appropriated carrying capacities' resulting from international trade agreements and the activities of transnational corporations.

With the economy continually at the centre of media and political attention, its apparent absence from the SSP may appear curious and invite criticism. It was a deliberate decision to subsume economic factors within the wider sociopolitical context, expressing a belief that 'the economy' should be a major means to an end, that is, sustainability, not an end in itself. The recent history of prices and financial transactions are implicit in the interactions among SERF variables, but SERF itself was conceived as an alternative to price-driven, self-equilibrating econometric models, as noted in Chapter 2. As discussed in Chapters 5 and 6, we recognize that the implementation of policies to move Canada toward sustainability would have to make use of appropriate market incentives and other economic instruments.

The scenario implicitly assumed that social equity could be maintained or enhanced while moving toward sustainability. But it did not really suggest how, beyond references to a guaranteed annual income, work sharing, and an acknowledgment of the 'informal' sector. This assumption might look naïve to people who call attention to the worsening situation in Canada and elsewhere, caught up with global competitiveness, employment downsizing across most sectors, and reduced commitment to social policies. While distributional issues are critical to achieving sustainability, with no mention of how vulnerable groups in particular might obtain justice and equity, the scenario appears to have just abandoned the issue. It was more a matter of not knowing what else to suggest.

Inclusion of a guaranteed annual income in the scenario invites criticism from political conservatives. The idea becomes a focal point for conflicting beliefs about the essence of human nature, and is interesting from that point alone. But this issue is moving into the political arena much faster than we had anticipated, in part because the increasing displacement of human labour by technological changes is impacting on the middle classes. Growing economic disparities and the potential for social unrest are inviting debate about alternatives, and to the cry of 'Where will the money come from?' the idea of a 'bit tax' on computerized transactions is being raised. Our scenario is already rather tame in its assumptions about changes in communications technologies and issues about the future of work.

Institutional arrangements for governance in a sustainable society need to receive more attention. The scenario's depiction of the devolution of responsibilities to the community level, but guided by the maintenance of strong national standards, is partly consistent with how governments are already responding to fiscal non-sustainability. Issues of how to develop effective multistakeholder consultations and partnership arrangements were not raised by the scenario, though they are arising now as alternatives are being sought to top-down administration by government. More contentious issues about property regimes for sustainability, and corporate governance for social accountability, are emerging sooner than we had anticipated. We did not address them, but our analysis supports the view that we need to confront these issues squarely.

The scenario was our social construction of a possible reality. By taking the time to articulate some underlying beliefs and values, and to work through their implications on SERF, perhaps we did not devote enough attention to describing our constructs of ecosystems and social systems. The environment/ecosystem components of the scenario were drawn implicitly from three 'schools' of ecology, that is, stress-response analyses, trophic dynamics, and landscape ecology. Social institutions, or 'structures,' were generally interpreted to mean the recurring patterns of human behaviour, created and re-created by the actions of people following different rule systems encoded in laws or customs. From this perspective, 'institutions' may be conceived as rule systems, with or without special organizations to administer them. The challenge, then, is to identify the rule systems, or individual rules, that support sustainability, and those that impede it.

Elements of complex systems theory entered into some of our thinking about sustainability, especially our recognition that the essence of sustainability is to maintain the capacity for natural renewal and continued evolution in ecosystems, and the capacity for innovation and creativity in social systems. All else are preferences for systems exhibiting different attributes, and there are elements of choices that can be made among them. The admonition about avoiding catastrophic change becomes an issue of scale, with the implied strategy of learning how best to avoid the positive feedback processes operating within both ecosystems and social systems that lead to tragic collapses of regional ecosystems and economies.

Complex systems theory should also have challenged us to think more consciously about issues of spatial and temporal scale. It posits a 'social construct' of nested sets of systems within systems, with each system exhibiting self-organizational capacities and emergent properties. For example, sustainability for any one level in such a hierarchy might be at the expense of sustainability at other higher or lower levels. With Canada being the entity or system level for which sustainability is sought, regions and communities may have to change in ways interpreted to mean that they are no longer sustainable. The metaphor of the 'ecological footprint' indicates that Canada's apparent sustainability is at the expense of 'appropriated carrying capacity' elsewhere in the world.

On a more jocular note, a pointed question can be posed to the SSP. Would we really want to live in the kind of world envisioned by the scenario? Obviously, it would depend on the alternatives. But the preference for a more community-oriented, localized management and use of resources and an enhanced self-reliance reflects one kind of conservatism. Intergenerational preferences could vary. The implied constraints on extended travel, for example, may not sit well with young adults. In order to enjoy a few drinks or to 'light up,' they would probably have to be furtive to avoid the scowls of the abstainer vegetarian elders sharing their multiple-generation living compounds. *Plus ça change...*

Criticisms have been, and rightly will be, directed to our assumptions, preferences, and outcomes. This is as was intended. If the SSP fosters further serious and thoughtful debate on the issues raised, then we would see this as an important accomplishment. The sterility of so much political debate and media commentary in recent years makes this serious discussion more important than we once thought. We urge others to create their own scenarios, and to make explicit their assumptions and beliefs.[1]

Notes

1 To that end, recent work at the Sustainable Development Research Institute has focused on the development of a user-friendly sustainability computer game, 'QUEST,' which will allow users to construct and evaluate their own scenarios of a sustainable future. The current version of QUEST, scheduled for completion in 1997, operates at the spatial scale of the Lower Fraser Basin in British Columbia; we hope to create versions applicable to other regions over the next several years.

Appendix A:
Project Participants and Papers

Project Participants
John B. Robinson, Sustainable Development Research Institute, University of British Columbia, Principal Investigator

George Francis, Department of Environment and Resource Studies, University of Waterloo, Co-investigator

Sally Lerner, Department of Environment and Resource Studies, University of Waterloo, Co-investigator

Russel Legge, Department of Religious Studies, University of Waterloo, Co-investigator

D. Scott Slocombe, Department of Geography and Environmental Studies, Wilfrid Laurier University, Co-investigator

David Biggs, Sustainable Development Research Institute, University of British Columbia, Graduate Student

Caroline Van Bers, Dovetail Consulting, Vancouver, BC, Project Manager

Other Researchers
Robert Gibson, Department of Environment and Resource Studies, University of Waterloo; areas of interest: decision-making and social organization, politics of the environment

James Kay, Department of Environment and Resource Studies, University of Waterloo; area of interest: technological change

Robbie Keith, Department of Environment and Resource Studies, University of Waterloo; areas of interest: minerals, agriculture

Bruce Mitchell, Department of Geography, University of Waterloo; areas of interest: fisheries, water

George Priddle, Department of Environment and Resource Studies, University of Waterloo; areas of interest: parks, wilderness areas

James Robinson, Department of Environment and Resource Studies, University of Waterloo; area of interest: water use and management

Advisory Committee
François Bregha, Resource Futures, International, Ottawa

David Brooks, Environment and Natural Resource Management, International Development Research Centre, Ottawa

Hélène Connor-Lajambe, Energy 21, Paris
Arthur Cordell, Industry Canada, Ottawa
Anthony Dorcey, Westwater Research Centre, University of British Columbia
Peter Duinker, School of Forestry, Lakehead University
Phil Elder, Faculty of Environmental Design, University of Calgary
Iris Fitzpatrick-Martin, GAMMA Institute, Montreal
Trevor Hancock, public health consultant, Kleinburg, Ontario
Stuart Hill, School of Social Ecology, University of Western Sydney, Australia
Robert Hoffman, ROBBERT Associates, Ottawa
John Hollins, Environmental Conservation Service, Environment Canada, Ottawa
Susan Holtz, National Round Table on the Environment and the Economy, Ottawa
Ray Jackson, Ottawa
Michael Keating, environmental writer and consultant, Toronto
James MacNeill, Institute for Research on Public Policy, Ottawa
Ted Manning, Centre for a Sustainable Future, Ottawa
Robert Paelhke, Department of Political Science, Trent University
Henry Regier, Institute for Environmental Studies, University of Toronto
William Ross, Faculty of Environmental Design, University of Calgary
David Runnalls, Institute for Research on Public Policy, Ottawa
Ted Schrecker, Westminster Institute for Ethics and Human Values, London, Ontario
Peter Timmerman, International Federation of Institutes of Advanced Study, Toronto
Ralph Torrie, Torrie-Smith Associates, Ottawa

Published Papers

Robinson, J.B. Forthcoming. 'Exploring a Sustainable Future for Canada.' In *Global Environmental Risk*. Ed. Roger Kasperson. Tokyo: UNU Press

Robinson, J.B., G. Francis, R. Legge, and S. Lerner. 1990. 'Designing a Sustainable Society: Values, Principles and Definition.' *Alternatives* 17 (2):36-46

Robinson, J.B., and C. Van Bers. 1991. 'Exploring a Sustainable Future for Canada: The Next Step in the Conserver Society Discussion.' *Studies for the Twenty-First Century*. Ed. M. Garrett, G. Barney, J. Hommel, and K. Barney. Arlington, VA: Institute for 21st Century Studies

Slocombe, D.S., and C. Van Bers. 1991. 'Seeking Substance in Sustainable Development.' *Journal of Environmental Education* 23 (1):11-18

Van Bers, C., and J.B. Robinson. 1993. 'Farming in 2031: A Scenario of Sustainable Agriculture in Canada.' *Journal of Sustainable Agriculture* 4 (1):41-65

Working Papers

Robinson, J.B., G. Francis, R. Legge, and S. Lerner. 1990. 'Defining a Sustainable Society: Values, Principles and Definitions.' SSP Working Paper #1, Department of Environment and Resource Studies, University of Waterloo, Waterloo, July

Slocombe, D.S., and C. Van Bers. 1991. 'Ecological Design Criteria for a Sustainable Canadian Society.' SSP Working Paper #2, Department of Environment and Resource Studies, University of Waterloo, Waterloo, March

Lerner, S. 1991. 'Socio-Political Design Criteria for a Sustainable Canadian Society.' SSP Working Paper #3, Department of Environment and Resource Studies, University of Waterloo, Waterloo, July

Robinson, J.B., G. Francis, R. Legge, and S. Lerner. 1992. 'Canada as a Sustainable Society: Environmental and Socio-Political Dimensions.' SSP Working Paper #4, Department of Environment and Resource Studies, University of Waterloo, Waterloo, September

Biggs, D., D. McFarlane, L. Kalbfleisch, C. Van Bers, S. Lerner, and J.B. Robinson. 1992. 'Canada in 2030: The Sustainability Scenario.' SSP Working Paper #5, Sustainable Development Research Institute, University of British Columbia, October

Biggs, D. 1994 'Exploring a Future Sustainable Energy Scenario for Canada between 1990 and 1030.' SSP Working Paper #6, Sustainable Development Research Institute, University of British Columbia, Vancouver, February

Van Bers, C., and J.B. Robinson. 1992. 'Farming in 2031: A Scenario of Sustainable Agriculture in Canada.' SSP Working Paper #7, Department of Environment and Resource Studies, University of Waterloo, Waterloo, November

McFarlane, D. 1993. 'Health Care in Transition: Working Our Way towards a Healthier Future Canadian Society in 2030.' SSP Working Paper #8, Geography Department, University of Waterloo, Waterloo, October

SSP Presentations

Robinson, J.B. 1993. 'The Sustainable Society Project.' Invited paper presented at The Future of the Global Environment: The Role of Canadian and Japanese Science and Technology Symposium, Ottawa, 19 October. Canadian Global Change Program Incidental Report Series, Report IR94-2, ISSN 1194-6481, June 1994

Robinson, J.B., and C. Van Bers. 1991. 'Exploring a Sustainable Future for Canada.' Paper presented at The Social Implications of Technology conference of the Toronto chapter of the IEEE Society, Toronto, 21-22 June

Appendix B: The Socio-Economic Resource Framework

David Biggs

The Socio-Economic Resource Framework (SERF) is a large Vax Mainframe-based simulation framework which allows for the development and evaluation of long-range scenarios of Canada. This appendix provides an overview of the scope, approach, and structure of SERF.

Scale and Scope

SERF is large in scale and rich in compositional detail. It consists of about 2,000 multidimensional variables that are equivalent to about 400,000 time series. These variables represent vintaged stocks such as houses, infrastructure facilities, consumer durables, and vehicles, as well as the 500 goods and 200 activities of the input-output tables for Canada.

Most components in SERF are national in scope and recognize no spatial distribution of activities within these boundaries. Only two of SERF's subcomponents – population and dwellings – are implemented at the provincial level.

The time horizon for SERF is relatively long – forty years. This is required to analyze decisions that must be made in the near future involving investments in facilities that have useful lives of twenty or more years.

The time step of one year is common to all components in SERF. Short-term phenomena such as seasonal changes or business cycles are not addressed.

The Modelling Approach

SERF is an implementation of the design approach to modelling developed between 1974 and 1986 at the Structural Analysis Division of Statistics Canada. The term 'design approach' refers to both the act of designing futures through simulation and the use of design information to build the models used for such simulation. The models, the simulation framework underlying those models, and the user combine to produce a way of exploring alternative futures.

The approach incorporates principles of general systems theory and control theory. Models of the physical transformation processes of the socioeconomic system, such as demographics, consumption, production, and resource extraction, are linked together, but with the main control variables accessible to the user of the framework. The user, who may be an individual or an individual combined with a model of decision processes, is an integral part of the system, exploring alternative futures, providing novelty and change through scenarios (which are specifications of the control settings over the time period of the simulation), and learning from resolving inconsistencies through repeated simulations.

To make the user so central requires a separation between the decision processes that control the system and the physical transformation processes that underlie it. The former processes, which consist of human decisions and behavioural relationships, are controlled directly by the model user, while the latter, consisting of physical, technological, and biological relationships only, are expressed in the form of models. This separation requires certain restrictions on the simulation framework and on the constituent models of physical process.

To give the user control, the simulation framework must impose no optimization or equilibrium conditions, and it must also make explicit the tensions between supply and demand which the user may choose to resolve. That is, the scenarios that result from use of simulation framework are driven by user-supplied inputs in terms of decision processes, not by endogenous behavioural relationships defined in terms of physical processes. The physical process models impose no such relationship but merely indicate the physical and technological implications of the input assumptions.

With no optimization, simulations can produce physically inconsistent or socially unacceptable futures. An example of the former would be unavailability of the resources required to meet demands in the scenario; an example of the latter might be high levels of unemployment. In the language of the design approach, these inconsistencies are called 'tensions,' and it is central to the approach that tensions be reported to the user and resolved through changed input assumptions and repeated simulations. It is through this process that understanding of the socioeconomic system grows, and only when a scenario produces a simulation which satisfies both physical constraints and the social constraints imposed by the user is it considered consistent or balanced. A balanced scenario forms the basis for variations which establish bounds on policy options and the sensitivity of the system to policy changes. In this sense the simulation framework serves as a basis for testing the feasibility of and revealing some of the impacts of alternative user-defined views of the future.

Model Structure

In modelling physical aspects of the Canadian socioeconomic system, SERF provides a disaggregated representation of stock/flow relationships and vintage, or age, characteristics of the energy, material, and people in Canadian society. These relationships and characteristics are maintained in forty-three separate submodels (see Figure B.1) and over 1,700 multidimensional variables that are based on the extensive Statistics Canada database. As shown in Figure B.2, these models describe energy, labour, and material flows in four major components: demography (population, household formation, and labour force), consumption (housing, consumer goods, health care, education, trans-

Figure B.1

Conceptual hierarchy of SERF

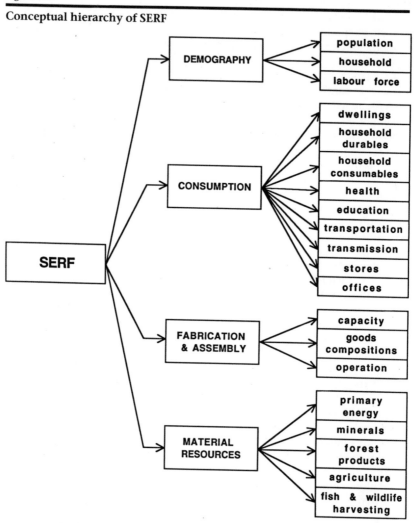

portation, offices, communications, and retail trade), production (detailed input-output and capacity models), and resource extraction (agriculture, forestry, primary energy, minerals, and wildlife harvesting). The disaggregated and comprehensive nature of SERF allows the user to undertake detailed analyses of changing efficiencies, technological substitutions, labour productivity, and so on. The user supplies input scenarios in the form of assumed future values for the approximately 1,700 multidimensional SERF variables, and SERF combines these time series inputs into integrated scenarios and assesses the physical consistency, over time, of the resultant overall scenario of the evolution of Canadian society.

Figure B.2

SERF dependency diagram

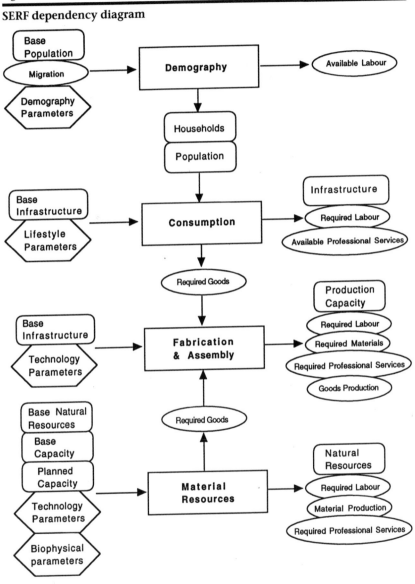

The demography component represents the basic demographic processes of population dynamics, household formation, and labour force participation. It keeps track of population by age and sex. Spatial distribution is represented at the provincial level. The user controls variables that reflect decisions about fertility, migration, family formation, and labour force participation.

The consumption component represents the infrastructure or stocks of goods that yield services required by human society. In general, it calculates the flows of goods, energy, and labour that are required to put infrastructure in place and to operate it. The consumption component keeps track of dwellings, consumer goods, hospitals, schools, motor vehicles, highways, airports, railroads, port facilities, hotels, restaurants, department stores, banks, and recreational and cultural facilities. The consumption models keep track of the goods, raw materials, and labour required for these activities.

The consumption component does not correspond to 'consumption' according to national accounting definitions. The emphasis here is on the availability of stock, not on the measurement of the value of the flow. The control variables reflect decisions to put infrastructure in place, to change its operating characteristics, and to consume the goods and services provided by that infrastructure. By having the consumption component follow the demography component, these decisions can take the form of parameters that reflect accessibility of intensity of use per capita or per family. In this way consistency between population and infrastructure can be assured. The models in the consumption component are dominated by stock/flow accounting and, as such, are analogous to population accounting.

The fabrication and assembly component represents the processes that transform materials and primary energy into finished goods that are required by both the consumption and material resources components. It keeps track of the stock of productive capacity and the exchange of domestically produced goods and materials for those produced in other countries. An input-output model that distinguishes 200 sectors and 500 commodities is used to represent goods production. The fabrication and assembly component requires raw materials and primary energy from the material resources component, professional services from the consumption component, and labour from the demographic component.

The material resources component represents the activities of exploration, extraction, and refining of non-renewable resources – coal, oil, gas, metals, and non-metallic minerals – and those of managing and harvesting renewable resources – livestock, crops, forest products, and fish. It also includes the generation of electricity from hydro sites. This component shares many design features with the production component. The boundary between materials resources, fabrication, and assembly components is arbitrary and is chosen to portray the tension between the availability of raw materials and the requirements for them. The material resources component also keeps track of the availability of resources to the extent that it is known. Additions to the stock of 'producible reserves' of non-renewable resources are the result of exploration activity; withdrawals, the result of extraction activity. For renewable resources such as forestry, additions are the result of growth which may be enhanced by forest management activities, but may be retarded by pollutants that are the result of human activity.

Tensions

The information flows among the four components of SERF, or the dependency structure, have been designed to highlight three sets of tensions. These include tensions between

- the availability of labour in the demographic component and the use of labour in the consumption, the material resources, and the fabrication and assembly components
- the availability of materials and primary energy in the material resources component and their use in both the fabrication and assembly and material resources components
- the availability of professional services in the consumption component and their use in the consumption, material resources, and fabrication and assembly components.

These are by no means the only tensions identifiable with SERF. For example, there is also tension

- in the exchange of domestically produced materials, goods, and services for those produced in other countries
- between the stock of productive capacity and its utilization
- between exploration activity which yields producible reserves and extraction from these reserves.

In the absence of models of decision processes in the model structure, tension resolution is achieved by user intervention. In other words, SERF does not determine how society will respond to these tensions. They are reported to the user who then must decide how to resolve them. In an iterative process, the user examines the tensions, decides on a scenario strategy to resolve them, then reruns the model with these new scenario assumptions. This process continues until the tensions are resolved to the satisfaction of the user. In this way, the user is playing the role of society in the modelling process rather than that role being simulated by the model itself. This allows the user to employ intuition and creativity in simulating the behaviour of society in dynamic and complex future scenarios.

Further Information

The SERF model is no longer operational. However, modelling efforts using the design approach and the backcasting technique continue at the Sustainable Development Research Institute. These efforts have focused on the development of QUEST, a user-friendly computer game/model that operates at the regional spatial scale of the Lower Fraser Basin. For more detailed information on SERF or the design approach, consult the following sources:

Gault, F., K.E. Hamilton, R.B Hoffman, and B.C. McInnis. 1987. 'The Design Approach to Socio-Economic Forecasting.' *Futures* 19 (1):3-25

Hoffman, R. 1986. 'Overview of the Socio-Economic Resource Framework (SERF).' Working Paper 86-03-01, Structural Analysis Division, Statistics Canada, Ottawa

Hoffman, R., and B. McInnis. 1988. 'The Evolution of Socio-Economic Modelling in Canada.' *Technological Forecasting and Social Change* 33 (4):311-24

Contributors

David Biggs is a graduate student in Resource Management and Environmental Studies at the University of British Columbia.

George Francis is a professor in the Department of Environment and Resource Studies at the University of Waterloo.

Russel Legge is an associate professor in the Department of Religious Studies at the University of Waterloo.

Sally Lerner is a professor in the Department of Environment and Resource Studies at the University of Waterloo.

John B. Robinson is the director of the Sustainable Development Research Institute and a professor in the Department of Geography, both at the University of British Columbia.

D. Scott Slocombe is an associate professor in the Department of Geography and Environmental Studies at Wilfrid Laurier University in Waterloo, Ontario.

Jon Tinker works as a consultant and writer. He is a senior associate with the Sustainable Development Research Institute at the University of British Columbia.

Caroline Van Bers is a senior associate with Dovetail Consulting in Vancouver, British Columbia.

Index

The Sustainable Development Research Institute (SDRI) at UBC was established in 1991 to initiate and contribute to interdisciplinary research on linkages between the environment, the economy, and society. SDRI is a vehicle for the development and coordination of sustainable development initiatives on campus and a process for encouraging interdisciplinary collaboration among the faculty departments and centres at UBC and other provincial institutes undertaking environmental research. It serves as a regional link with government, the private sector, and other research institutions engaged in sustainable development research across Canada.

SDRI's orientation is focused on producing applied, policy-relevant, and interdisciplinary research. We believe that there is an increasing need to move away from predicting the negative impacts of environmentally sustainable behaviour. Instead, our country must move in the direction of articulating ecologically and socioeconomically desirable futures, and determine what needs to be done to make these futures happen. The environmental agenda encompasses much more than staving off ecological disaster; it involves trying to create a society that is sustainable, in all manifold dimensions.

SDRI's objectives are:

- to undertake interdisciplinary research on the environmental, economic, social, and cultural changes necessary for sustainable development and the ways in which these changes can be implemented
- to contribute and disseminate useful knowledge about these changes to local, national, and international communities
- to foster links and conduct collaborative research with government, non-government organizations, and the private sector, regionally and internationally.
- to encourage academic opportunities at UBC for interdisciplinary and transdisciplinary studies related to sustainable development.

Set in ITC Stone by Typeworks

Printed and bound in Canada by Friesens

Copy-editor: Dallas Harrison

Proofreader: Maureen Nicholson

Indexer: Annette Lorek